Thinking Time Geography

Time-geography is a mode of thinking that helps in the understanding of change in society, the wider context and ecological consequences of human actions. This book presents its assumptions, concepts and methods, and example applications.

The intellectual path of the Swedish geographer Torsten Hägerstrand is a key foundation for this book. His research contributions are shown in the context of the urbanization of Sweden, involvement in the emerging planning sector and empirical studies on Swedish emigration. Migration and innovation diffusion studies paved the way for prioritizing time and space dimensions and recognizing time and space as *unity*. From these insights time-geography grew. This book includes the ontological grounds and concepts as well as the specific notation system of time-geography – a visual language for interdisciplinary research and communication. Applications are divided into themes: urban and regional planning; transportation and communication; organization of production and work; everyday life, wellbeing and household division of labor; and ecological sustainability – time-geographic studies on resource use.

This book looks at the outlook for this developing branch of research and the future application of time-geography to societal and academic contexts. Its interdisciplinary nature will be appealing to postgraduates and researchers who are interested in human geography, urban and regional planning and sociology.

Kajsa Ellegård is Professor in Technology and Social Change, Linköping University, Sweden.

Routledge Studies in Human Geography

This series provides a forum for innovative, vibrant, and critical debate within Human Geography. Titles will reflect the wealth of research which is taking place in this diverse and ever-expanding field. Contributions will be drawn from the main sub-disciplines and from innovative areas of work which have no particular sub-disciplinary allegiances.

For more information about this series, please visit: www.routledge.com/Routledge-Studies-in-Human-Geography/book-series/SE0514

Thinking Time Geography
Concepts, Methods and Applications

Kajsa Ellegård

Routledge
Taylor & Francis Group
LONDON AND NEW YORK

First published 2019
by Routledge
2 Park Square, Milton Park, Abingdon, Oxon OX14 4RN

and by Routledge
52 Vanderbilt Avenue, New York, NY 10017, USA

First issued in paperback 2020

Routledge is an imprint of the Taylor & Francis Group, an informa business

British Library Cataloguing-in-Publication Data
A catalogue record for this book is available from the British Library

Library of Congress Cataloging-in-Publication Data
A catalog record has been requested for this book

ISBN 13: 978-0-367-58586-0 (pbk)
ISBN 13: 978-1-138-57379-6 (hbk)

Typeset in Times New Roman
by Swales & Willis Ltd, Exeter, Devon, UK

Contents

Figures

Tables

Preface

This book is the result of many years of living with a time-geographic mindset. It was founded when I, as a human geography student, read the most intriguing and unique national investigation I had experienced until then. Written by Professor Torsten Hägerstrand, it demonstrates how to overcome the cleavage between macro-level aggregate averages and micro-level everyday life as lived by people. This was an eye-opener, and when I in the autumn of 1975 was asked to serve as a research assistant in a research project about the future public transportation systems in Scandinavian cities, led by Bo Lenntorp and Torsten Hägerstrand, I could not resist. I was involved in the time-geography research group in Lund University, which grounded me for my engagement in developing the time-geographic approach in the coming years in studies of a wide flora of empirical fields. The studies concern, for example, milking cows; the development of the dairy industry; production and work organization in the automobile industry; the integration of teachers' work tasks in their daily life activities; and energy use in households. I have also developed a time-geographic diary method where the activity sequence of the diarist is put in the context of places visited, persons encountered, appliances used and subjective experiences of the activities.

This book is my summary of the time-geographic approach, its development and ontological grounds, concepts, methods and notation system. From the PhD courses in time-geography that I have held over the years, and from the yearly seminars in the Scandinavian time-geography network of researchers, I have experienced the lack of a book which presents the approach as a whole. Here is my contribution to the research community interested in getting deeper into time-geography. I hope this book will inspire coming generations of researchers interested in developing new knowledge about how human activities in the time-space influence life in the environmental context.

When I went out to fetch the cows for milking in the summer afternoons in 1975, I could not imagine what influence this experience would have on what I have done in my life. Obviously, there were more opportunities in the future than I had in my mind. My individual path as a researcher can be traced back to the coupling between milking cows and the reading of Hägerstrand's 1972 work. I hope that young researchers reading this book will find possible couplings between the time-geographic thinking and their research interests.

Last but not least, I want to say that the inestimable harsh and honest critique, combined with invaluable encouragement to go further in the work on and in time- geography, given by my colleague and friend Bo Lenntorp has strengthened my paving the time-geographic path. It has been a pleasure even to be criticized!

13 May 2018
Kajsa Ellegård

Acknowledgements

The project that made this book possible is funded by Riksbankens Jubileumsfond, in the RJ sabbatical program. Without that support the book would have remained an idea. The project made it possible for me to visit many scholars engaged in time-geographic research. I would especially like to thank my hosts at the universities visited; Professor Andrew S. Harvey, St. Mary's University, Halifax, Canada; Professor William (Bill) Michelson, University of Toronto, Canada; Associate Professor Dana Anaby, McGill University, Montreal, Canada; Professor Yanwei Chai, Peking University, Beijing, China; Associate Professor Tim Schwanen, Oxford University, Oxford, England; Professor Masago Fujiwara, University of Shimane, Hamada, Japan; Professor Kohei Okamoto, Nagoya University, Nagoya, Japan; Professor Martin Dijst, Utrecht University, the Netherlands, now at Luxembourg Institute of Socio-Economic Research, Luxembourg; Associate Professor Eva Magnus, Norwegian University of Science and Technology (NTNU), Trondheim, Norway; Professor Mei-Po Kwan, University of Illinois, Urbana-Champaign, USA; Professor Harvey J. Miller, Ohio State University, Columbus, USA; and Professor Shih-Lung Shaw, University of Tennessee, Knoxville, USA. I also want to thank all of their colleagues for attending the seminars and lectures I gave, and for good discussions at all these places.

I want to thank three Swedish researchers for sharing their different experiences and perspectives on the spread of time-geography. On one hand, the human geographers Professor Bo Lenntorp and Associate Professor Solveig Mårtensson, who both were engaged in Hägerstrand's research group from the 1960s, and, on the other hand, occupational therapist Ulla Kroksmark, who introduced time-geography to the occupational science community. As a member of the research group Technology, Everyday Life, Society (TEVS) at my department, Technology and Social Change, Linköping University, Sweden, I have had the privilege of presenting drafts of chapters of this book and the critical discussions have helped me to improve the ideas and the text.

My frankest critic, Bo Lenntorp, has read everything and pointed out my weaknesses and given support during this process – thank you for this and for our collaboration since the mid 1970s! In the end, the mistakes and misinterpretations that remain are solely my own.

Part I

Introducing the time-geographic approach

1 Introduction – origin and societal context

Capturing societal change

On a journey from one big city to another in modern Sweden, the trip goes through rural areas displaying a great variation in the scenery, with forests, lakes, villages, farms, fields with crops and not many people. Here and there, at some distance from modern settlements, a desolate farmhouse appears, with a broken roof beam, windows and door fallen apart and a big tree growing too close to the wall; it resembles a still-life arrangement. However, such farmhouses enclose a lot of human activity. Once, people worked hard to construct and maintain them, and they must have been proud when the new house was completed and the family could establish a life there, subsequently sustaining themselves by working on the small farm. The modern traveler passing by gets a quick glimpse of the material remains of this life, but will probably not reflect over what is embedded in it. What once was an important resource for sustenance of a whole family is in the modern society of little worth, if any. What was a resource and a source of pride is turned into something to pity or not notice at all. The traveler also passes relics of small-scale factories, which were established in the late 19th and early 20th centuries, close to raw materials and water power. These too have met a similar destiny as the small farms, resulting in factory lay-offs and people moving into larger cities.

The desolate farmhouses and factory buildings indicate fundamental societal change over time, which gives rise to reflections about what is a resource, and what activities do people engage in over time at various geographical settlements. It underlines the importance of taking time and place into consideration when reflecting over and investigating human activity in current societies.

The time-geographic approach provides conceptual tools and a notation system useful for investigating processes of societal change. It helps in analyzing how one and the same need is satisfied differently depending on where, when and by whom the activities are performed. It has to do with variations in available resources, in terms of knowledge, technologies and tools, and the opportunities for people to arrange the resources so that they are within reach when needed to perform activities aimed at achieving goals and satisfying needs. The latter concerns couplings in time and space, which is a main issue for time-geography developed by human geographer Torsten Hägerstrand.

Successful performance of activities to achieve goals creates couplings in time and space between the involved persons on one hand, and these persons and the tools and other resources needed on the other. Then, the time-space location of people and resources is essential. When considering couplings in time and space the individual's dependence on other individuals is underlined, which helps reveal what might hinder, or facilitate, the achievement of individual and organizational goals. The two seemingly simple dimensions of time and space help sort out things that otherwise might be perceived as entangled and not subject to a coherent logic.[1]

The time-space couplings in different contexts

The daily activities in modern households are the results of coordination between the household members concerning engagement in different projects. Households are also dependent on the schedules and locations of workplaces, and the supply of services, schools and day care centers. Increasing geographic centralization of service supply makes people more dependent on transportation. This section shows how the organization of serving a meal in two types of societies puts very different claims on household members' time and coordination of their activities.

Fundamental long-term changes in societal conditions for production and consumption are revealed by a closer look at serving a meal in a household in an agricultural and an industrial society respectively. Generally, meals relate to the inevitable human need for food, include a basic relationship to nature for ingredients and relate to how possible couplings in time and space are arranged between people and various kinds of resources.

In an agricultural society, the serving of a meal in a household requires fundamentally different activities from household members compared to a household in an industrial society. It concerns different time perspectives as well as differences regarding the production means and organization, geographical location of production relative to consumption, organizations' specialization, size of production units, use of resources and handling of waste[2] and need for transportation, and what is relevant knowledge. An outcome is that the total time household members spend on serving a meal today is relatively limited compared to that of those in the agricultural society, and so is the length of the total time period within which the household members perform the necessary activities for it.

In today's society, a household can get a meal in many ways, all based on production of ingredients in specialized organizations. They can use ingredients bought earlier or go to the grocery shop to buy what is needed, and they can go to a restaurant providing meals cooked by a professional chef. In this case, not only the production of the ingredients but also the serving of the meal is decoupled from the household members and put onto professionals.

In contrast, in an agricultural society, a meal in a farm household was the outcome of the combined work of household members engaged in the many different kinds of activities and projects at the farm during the whole year. They worked in the field, in the meadow, in the cowshed and stable and workshop. Household members performed activities for producing food in a coordinated sequence over the year, which finally turned the combination of work efforts and materials into the ingredients necessary for cooking food. Then, to realize the meal in the agricultural society the farm household members had to master and coordinate activities related to a whole body of consistent knowledge, which today is regarded as fundamentally different kinds of knowledge fields.[3] The whole and varied body of knowledge mastered by the households in the agricultural society is today differentiated and replaced by industrial activities. Instead, different industries specialize in food production.

In the Scandinavian agricultural society, production and consumption activities were performed in close vicinity and the production was immediately dependent on the local climate, weather, soil quality and the knowledge among the household members and their work in the production. It took a long time to produce the raw materials needed for serving a meal, and the agricultural households covered the whole process from seed to bread. There were limited needs for long-distance transport of raw materials. This is what Hägerstrand called the *vertically linked society* (Hägerstrand 1970b); later he used the term *short-distance society* for it (Hägerstrand 1988).

The production of raw materials and ingredients of a meal for the modern household, on the contrary, is decoupled from most of the activities performed by the household members, not only in a material sense but also as regards knowledge about the production. Thereby, the time spent by the modern household members on activities necessary for serving a meal is much shorter than in the agricultural society, and it is also less directly coupled to the local agricultural conditions.

Today, the ingredients of the meal lean heavily upon specialization and concentration of the food industry, where the production units and distribution systems are mainly organized according to large-scale production principles and economies of scale. Raw materials and semi-finished products for food production, then, are transported from different places in the world to food industries and further to grocery shops to be bought by members of households. Hägerstrand called a society with such a geographically spread location of activities and production units the *horizontally linked society* (Hägerstrand 1970b). Later on, he used the term *long-distance society* for this societal organization of production (Hägerstrand 1988). The many long-distance journeys are possible because of the relatively cheap supply of fuels, transportation methods with cooling equipment and a huge infrastructure for transportation. Specialization of food production meant decoupling it from households, which was a prerequisite for the development of the industrial society with its need for labor and for urbanization. Taken together, the processes have freed most people from the time-space couplings that bound them to the hard farming work activities, but instead they are dependent on food industries.

Of course, the horizontally linked society, just like the vertically linked society, depends on the local climate, weather and the fertility of the soil in the places where crops are grown and animals nourished. Thereby, the production of raw materials for food, at the very place where the crops are growing, is similar in the two types of societies, but the consequences differ. For example, in the vertically linked society extreme weather might lead to lack of food and starvation for the household, while in the modern society the urbanized customers might experience the extreme weather conditions solely via rising prices caused by decreased raw material supply.

Whether the society is mainly vertically or horizontally linked, the production of ingredients and the cooking of a meal requires coordination in time and space, both as regards people and material and immaterial resources. Time-geography provides tools for understanding such combinatory principles in time and space of relevance for production and consumption in the daily life of people in different kinds of societies.

Hägerstrand used the ongoing specialization and concentration in agricultural production to exemplify the long-term societal changes, from a vertically linked, short-distance society into a horizontally linked, long-distance society. Partly this emanates from his personal experience of the changing and rapidly urbanizing Sweden, with increasing mobility, specialization and geographical concentrations of industrial and service activities as well as people.

Torsten Hägerstrand: a short biography

Torsten Hägerstrand's biography is interlaced with the long-term changes in Sweden that fundamentally transformed the society. From the late 19th century and during the 20th century, Sweden went from being a poor agricultural society dominated by a rural population and high emigration figures into a modern industrialized welfare society, with an urbanized population and high immigration, which will be dealt with in Chapter 4. At the societal level, these processes influenced the location and organization of activities in production, administration and services, resulting in specialization, concentration and large-scale organizations. At the individual and household level, the process gave rise to developments such as those exemplified in the section above. Sweden is currently a rich postindustrial country with high material living standards and a general social welfare system recognized worldwide.

Torsten Hägerstand, born in 1916, lived his early life in the small community Torps Bruk, located in the forested county of Småland in Southern Sweden. The region was dominated by small farms and small factories built close to raw materials and water-power sources. He experienced the different living conditions among his friends, whose fathers worked in various occupations in the small-scale production industries, agriculture and services.

Hägerstrand's father was a schoolmaster and the family lived on the second floor of the schoolhouse. The schoolchildren's presence and absence in the building and outside on the schoolyard made the young Hägerstrand reflect on variations in daily rhythms depending on day of the week and time of the day, which he commented upon in *The Practice of Geography* (1983). Hence, the location of his home in the schoolhouse made him experience the difference in activities and social encounters at the same place in different time perspectives.

His father, the schoolmaster, underlined the importance of studies to Hägerstrand from his early childhood. The school subject *hembygdskunskap* (home area studies) was introduced in the Swedish primary school system in the early 20th century and his father eagerly taught it, not just to the pupils in class but also to the children in the Hägerstrand family. The home area studies subject mixed knowledge from different scientific disciplines, i.e. biology, history, archaeology, geography and ethnography, and made it relevant for the pupils by setting it into the context of the local community. The schoolchildren's own experiences of the nature and social life in their neighborhood was thereby related to general knowledge of the schoolbooks. In his teaching, the schoolmaster was inspired by the Swiss pedagogue Pestalozzi, who claimed that the schoolchildren should learn about the world from what is close to them and outwards. Hence, for example, the pupils learnt how to draw a map of their classroom, then the schoolyard, and after that larger areas were mapped, thereby emphasizing the importance of scaling the objects to be shown on a map (Hägerstrand 1983).

The co-existence of, and mutual interdependence and competition between, non-living things (material artifacts and things created by nature) and living entities in the schoolyard and its vicinity (a mix of flowers, trees, insects, human beings, wild and domesticated animals, and so on) fascinated the young Hägerstrand. This interest grounded him for thinking about what the co-existence of various kinds of species in a certain area meant for the living creatures as well as for the landscape. His interest was in finding out the conditions for the different phenomena and their existence in a context, rather than grouping species into fixed categories based on their similarity.

Hägerstrand entered Lund University as a student in the late 1930s, and early on he became puzzled by the dominating specialization in academia – which differed fundamentally from the knowledge development he experienced in school, based in the home area studies subject and the pedagogy of Pestalozzi. Later in life, as an established researcher he promoted the idea of transgressing the borders between scientific disciplines construed by humans.[4] After studies in several disciplines, among them ethnography, geography, archaeology and mathematics, he graduated and started his PhD studies in geography, a discipline that he found less specialized than most of the others. Initially his research concerned migration in a small Swedish parish during the 19th century and the emigration to America from there. The insights and conclusions from this study are presented in depth in Chapter 2. Thereafter, he studied diffusion of innovations in the same geographic area. The latter study resulted in Hägerstrand's PhD thesis, defended in 1953. From the studies of migration and innovations, his interest was directed to individuals' continuous movements in both time (continuous sequences of events) and space (location and transportation of material/physical entities). This theorization on both migration chains and innovation waves laid the ground for developing the thinking that was eventually presented as the time-geographic approach. The general time-geographical approach, with its ontology, concepts and notation system, is presented in chapters 2 and 3.

Hägerstrand was appointed Professor in Human Geography in 1957, in a period when the Swedish society was rapidly changing and the urbanization movement was strong. There was a demand for academic knowledge to understand the process and to steer societal development. In the mid 20th century policy makers at local, regional and national levels, some of them Hägerstrand's former students, engaged their former teacher as an expert and source of inspiration in their efforts to plan for the development of the welfare state. By then, there was a relatively small number of social science researchers in Sweden and many of them were engaged in constructing the welfare society.

This societal transformation, in combination with the conclusions from research on the importance of considering the time-space movements of individual phenomena, frames the development of time-geography. The societal transformation of Sweden and its influence on the development of time-geography is presented more in depth in Chapter 4.

Hägerstrand's time-geography was in the making when he, together with a number of university professors in human geography and economics, got a big grant from the research foundation Riksbankens Jubileumsfond[5] for a joint social science research project, "The urbanization process". For Hägerstrand this project meant that for the first time in his career he had an opportunity to form a research group, and three young researchers – two men, Bo Lenntorp and Tommy Carlstein, and one woman, Solveig Mårtensson – worked together with him in the Research Group for Process and System Analysis in Human Geography. The research group, informally called the time-geography research group, grew over the years. Hägerstrand was appointed to a personal professorship in 1971, financed by the Swedish Research Council for Humanities and Social Science. With the personal professorship, he had good opportunities to engage fully in the further development of the time-geographic approach, now with increasing emphasis on the ecological orientation of the approach (Lenntorp 2010). In parallel, the activities in the time-geography research group continued until Hägerstrand's retirement in 1982. Some researchers in the group eventually had positions in universities in Sweden and went on with their time-geographically inspired research.[6]

After a long period of work with the growing community of planners in the Swedish bureaucracy, Hägerstrand became increasingly disappointed over the lack of interest in practicing the concrete means and theoretical ideas that he presented and discussed at the meetings. In the mid 1970s he wrote:

> it is very easy to dream up blue-prints for new undertakings but very hard to imagine their fate and their consequences for other legitimate processes when put into practice. Perhaps the trouble is that thought does not encounter in its own world the constraints of space and time.
>
> (Hägerstrand 1976: 334)

In the quotation, the notion of time and space is underlined and the two are regarded as underestimated constraints for realizing ideas and plans. One main conclusion is that the time and space dimensions should be put to the fore in order to better understand what human actions mean for social development and sustainability.

Hägerstrand wrote and published frequently all through his life, and over the years his texts were published in a rich variety of journals, proceedings, books and edited volumes, and it is not an easy task to find them all.[7] During his last 10 years, he wrote the second book of his academic career, a comprehensive summary of his thinking, *Tillvaroväven*, which was posthumously published in 2009.[8] Torsten Hägerstrand died in 2004.

The following citation from Hägerstrand (1995) serves as an overall introduction to the time-geographic approach:

> Our actions leave traces in the physical world. We produce things and bring about states of a sort that nature does not shape on its own. Most traces have a short duration. Others lead to more lasting changes. In most cases there is a limited and comprehensible purpose behind the specific actions. In addition, most actions – probably all – have consequences which were not taken into account in the moment of action. It is quite easy to discover such unintended consequences in the immediate neighbourhood. Consequences with a wide reach and a slow course are more difficult to grasp.
>
> (Hägerstrand 1995: 35)

Notes

1 The sorting, of course, appears differently as the social context changes.
2 What is defined as waste also varies depending of what societal context is under analysis.
3 The whole body of knowledge mastered by the household members concerned the quality of the soil, animal health, building techniques, slaughter, milk production and conservation (e.g. butter and cheese), production of tools, preservation of food raw materials to make them usable over a long time, adjusting the meal to the seasonal variation of available raw materials, and so on.
4 He was one of the founding fathers, influential researchers from different disciplines and research areas, who actively worked for the realization of the Department of Thematic Studies in Linköping University in the late 1970s. The department started its operations in 1980 and is focused on interdisciplinary research in thematic research areas of societal importance (initially it considered Technology and Social Change, Water in Nature and Society and Communication; later Child Studies and Gender Studies were also established). See www.liu.se/tema.
5 The Swedish Riksbank celebrated its 300-year anniversary in 1966 by initiating a research foundation directed to research in humanities and social science, Riksbankens Jubileumsfond.
6 Among the group members who became professors were Bo Lenntorp at Stockholm University, Sture Öberg at Uppsala University, and Kajsa Ellegård, Tora Friberg and Stefan Anderberg at Linköping University.
7 Lenntorp (2004) contains references to the publications by Torsten Hägerstrand.
8 *Tillvaroväven* (The Fabric of Existence) was not ready when Hägerstrand died in 2004 and it was edited by Ellegård and Svedin. It was published by the research council Formas in 2009.

References

Hägerstrand, T. 1970a. What about people in regional science? *Regional Science Association Papers*, Vol. XXIV, pp. 7–21.

Hägerstrand, T. 1970b. *Urbaniseringen – stadsutveckling och regionala olikheter*. Lund, Sweden: C.W.K. Gleerup.

Hägerstrand, T. 1976. Geography and the study of interaction between nature and society. *Geoforum*, Vol. 7, pp. 329–344.

Hägerstrand, T. 1983. In search for the sources of concepts. In *The Practice of Geography*. A. Buttimer (ed.). Harlow: Longman Higher Education, pp. 238–256.

Hägerstrand, T. 1988. Krafter som format det svenska kulturlandskapet (1988). *Mark och vatten år 2010*. Bostadsdepartementet, Stockholm, pp. 16–55. In German: Die Kräfte, welche die Schwedische Kulturlandschaft formten (1989). *Münchener Geographische Hefte*, 62:15–59.

Hägerstrand, T. 1995. Action in the physical everyday world. In *Diffusing Geography. Essays for Peter Haggett*. A.D. Cliff, P.R. Gould, A.G. Hoare and N.J. Thrift (eds). Oxford: Blackwell, pp. 35–45.

Hägerstrand, T. 2009. *Tillvaroväven*. K. Ellegård and U. Svedin (eds). Stockholm, Sweden: Formas.

Lenntorp, B. 2004. Publications by Torsten Hägerstrand 1938–2004. *Geografiska Annaler. Series B Human Geography*, Vol. 86 B, No. 4.

Lenntorp, B. 2010. Torsten Hägerstrands världsbild – några tankar om dess utveckling. *Geografiska Notiser*, 2.

2   Ontological grounds

Simple and complicated

Hägerstrand (1985: 195) wrote about the time-geographic approach, "The approach is not in itself a theory. It is rather an ontological contribution preceding formation of theory." He was struggling with the problem of the very short step between "what is self-evident and extremely complicated" (1985: 195). When looking at the development of the time-geographic approach in retrospect it is clear that it takes its points of departure in phenomena that seem self-evident, and utilizes these phenomena to construct concepts that assist interpretation of complicated relations and appearances at the micro and macro level.

The concept of time-geography in itself implies an assumption that time and space[1] (or place) exist, and the hyphen indicates that the interplay between the two dimensions is considered. In this chapter, the ontological grounds of the time-geographic approach are presented. It concerns assumptions about time and space/place and presents the roots for combining these dimensions, based on an example from Hägerstrand's early research on migration in Asby, Sweden. The importance of the thorough empiric research laying the ground for the development of the basic ideas of the time-geographic approach is thereby underlined.

Time and space and the time-space

The German philosopher Kant separated time (when) from space (where) and said in some lectures at the university in Königsberg that handling the time dimension is the task of history, while geographers should handle the space dimension.[2] Such a stance creates problems both for historians and geographers. On one hand, in most historic texts there are geographic anchor places, since events discussed in historic research took place somewhere. On the other hand, problems emerged among geographers in the regional geography tradition[3] when they strived to delimit "natural regions" from the geographical form and factors like population size, degree of urbanization, industrial activities and agriculture, and raw materials. Delimitation was a difficult task because of changes related to such factors, and change relates, of course, to time.

One time-geographic assumption is that everything that happens has a geographical location. From this position, even an idea has its geographical location because it is tied to, for example, people or books, and they are located somewhere. The specific location of an idea or phenomenon at a place also reveals the material context wherein it was created at a specific time.[4] Consequently, from a time-geographic position, both geography and history should consider both time and space.

Shortcomings in understanding developments of the world may appear when either time or space is put aside. The time-geographical approach includes conceptual tools to follow changes in time and space, and there is a specific notation system to visualize such located processes. This notation system can be used to create a common ground and point of departure for communicating what has happened, is happening and might happen in the future regarding processes of interest. The conceptual tools and notation system are presented in Chapter 3. Here, the basic assumptions regarding time and space/place will be put to the fore.

About time

"[T]ime does not admit escape for the individual. (. . .) As long as he is alive at all, he has to pass every point on the time-scale" (Hägerstrand 1970: 10). This statement indicates some time-geographical claims related to time:

- Time is assumed to exist.
- Time is continuous and has a direction and a constant pace.[5]
- Now constitutes the continuous transformation of the future into the past.
- Time is regarded as a useful tool to study processes and change.
- Time can be measured.
- Time is the most equally distributed resource.

Most people have, presumably, thoughts about what kind of a phenomenon time is. There are philosophical theories concerning time and in everyday life people think about and perceive time in many different ways, including the idea that time might not exist. Whatever the answer is to the question of the existence of time, the time-geographic assumption is that time, as it can be measured by clocks and calendars, is a useful instrument for understanding and explaining the development and change of phenomena in society and nature.[6]

Based on the assumption that time exists and is measureable, time-geography ascribes it some objectivity. Despite this objectivity, of course, time can be *experienced* in a great variety of ways by different people. This means that there is a subjective dimension of time, which must be recognized. However, in the basic structure of the time-geographic visualizations, the time dimension is utilized in an objectivistic way in order to create a common ground for investigating processes over time, in a similar way as the conventional map is utilized to show the geographic location of places. Then, a time-geographical visualization can be used as a starting point for an individual reflecting on her subjective experiences of an event or process.[7]

The time-geographic assumptions that time has a direction and a constant pace imply that events can be anchored along the time dimension. This also implies sequences of events (they appear before, simultaneously or after one another), and in this perspective events are parts of larger and longer processes. Heracles, the Greek philosopher, argued for a processual meaning of time. According to Plato, Heracles recognized the transience of material things and that everything is steadily in motion:

> You cannot step twice into the same river; for fresh waters are ever flowing in upon you.
>
> (quoted in Russell 1961: 63)

The English philosopher Bertrand Russell wrote in his book *The History of Western Philosophy* that the thoughts behind this saying ascribed to Heracles are difficult to handle in science. Russell claims that

> Science, like philosophy, has sought to escape from the doctrine of perpetual flux by finding some permanent substratum amid changing phenomena.
>
> (Russell 1961: 65)

The time-geographic idea is to illustrate time as a linear dimension with one direction going from the past, via now, into the future. This implies that the same moment in time never returns. It resembles the saying denoted to Heracles, "fresh waters are ever floating in upon you". Time can be seen as the fresh waters, while the person's feet standing in the flowing stream of water, just by standing there, might be regarded as the now. Looking at time from such a double perspective, both as a flow (the river water) and as an apprehension of now (the feet standing in the river with the flowing water), is intriguing and shows that time is an ambiguous phenomenon.

In time-geography the ambiguity of time is approached by the foundations of the notation system. In the notation system time is regarded as a continuous dimension, as visualized by the Time axis in Figure 2.1. The continuous time dimension shall be read from below and up, since now is constantly moving upwards. In the abstract time-geographic visualization, now is described by a line (the now-line) that is constantly moving along the continuous time dimension.[8] Then, time is illustrated on the vertical y-axis in diagrams, which differs from how time is usually illustrated in diagrams, as discrete time points on the horizontal x-axis. The purpose of the deviation from the convention is that time should be re-thought and not be regarded from a habitual perspective.

The now-line fulfills two purposes: first, it shows that now constitutes the important distinction between future and past. Below the now-line is the past. What has happened is "frozen" and, consequently, activities that are performed in the past cannot be undone. What has happened, however, can be experienced

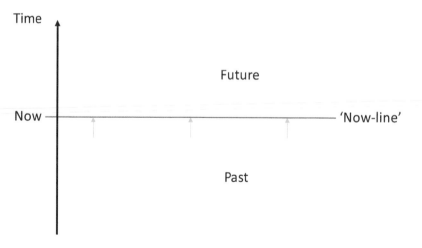

Figure 2.1 The now is located at one point in time along the Time axis, and now is
constantly moving upwards as time goes by. This movement is symbolized
with the small arrows under the "now-line". The now thereby transforms
future into past.

differently by different people, since they have different points of departure
and different goals for their actions.⁹ Then, even if a past situation never comes
back and cannot be undone, alternative *interpretations* of its appearance can be
made in retrospect.

Above the now-line is the future, in which there are several opportunities.
Planned projects might be fulfilled by people performing the necessary activities
as the now moves forwards (upwards) and transforms future into past. To achieve
the goals of their projects people must position themselves in the material environment so that the resources needed are within reach when they want to perform
the activities to achieve those goals.

Second, and related to now being the distinction between past and future, time
is continuous and has a constant pace (at the everyday level of humans). Thereby,
now is regarded as the constant transformation of future into past. This implies
that every action has to be taken now, in the very moment when future is transformed into past. Hence, now constitutes the only opportunity to act and thereby
to make a difference. This also means that the time dimension can be used to capture sequences of events, indicating relationships between before- and after-ness
in time when relating events to each other.

Since the now steadily moves along a one-directional time-dimension, there
is no cyclic time. However, there are events repeated every year, like ceremonial
occasions, according to the calendar, but such events are not cyclical in time; they
are just similar and appearing in sequence over time. The events are similar, not
the same, since each event happens at different points along the continuous time

axis (different nows). The calendar-based ceremonies are characterized by their appearance in a number of past nows and the planned (projected) appearance in the future, with some time distance in between. For example, even if New Year is an event every year it is of course not the same New Year repeated every year; instead it is different New Year-nows. Every new New Year-now might be organized in a similar way, year after year, but since now moves on, the same New Year never comes back.

As indicated above, time is regarded as continuous with a constant pace irrespective of human wants and wills. According to time-geographic ideas, the pace is constant even if a human being might experience time flying quickly or moving slowly. The continuous flow of time with its constant pace, then, cannot be halted. In time-geography, the lifetime of every human being as an indivisible unit (at the everyday level of thinking) is anchored on the historic time dimension from the moment of birth until death. The life experiences of two individuals, A and B, who are born with some decades between them, will consequently differ (see Figure 2.2). The specific differences will, among other things, depend on what happened in society during their lives. As a child, person A, born in 1940,

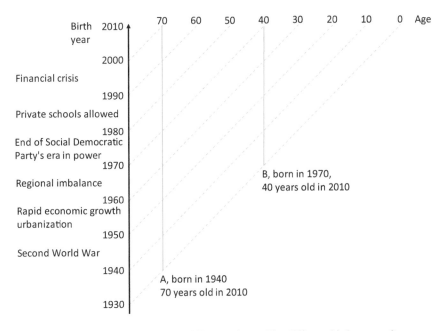

Figure 2.2 Influence of birth year on life experience. The different birth years of two people, A and B, reveal some differences in their life experiences as a result of what happened in society at various ages. A, born in 1940, experiences the Second World War from a Swedish perspective and also the rapid economic growth of the 1950s and 1960s, while B grows up in a period of new political power relations and financial crises. Figure based on the ideas presented in Hägerstrand (1972).

experienced the Second World War, while person B grew up during the wealthier years of the 1970s.

In a day perspective, time is the most equally distributed resource among people. Everybody has 24 hours of life every day, apart from the days of their birth and death, which are shorter. Also, everybody must live through all of the 24 hours per day as they appear in sequence. Therefore, at the individual level, time cannot be compressed, saved or omitted. All 24 hours must be filled with activity, even if the activity might be experienced as "doing nothing".

A human being is indivisible and regarded as a whole with a body that cannot be divided or truncated without loss of important functions or meanings. A similar reasoning can be used when considering human life as it appears in the course of time. The time flow passing during a human's life, or just during a day in a life, cannot be chopped into pieces without losing important meanings. For example, as time flows in a day perspective, a person has breakfast, lunch and dinner at different times of the day due to the biological need for a regular intake of food. The meals are eaten at three different times of the day and other activities are performed between the meals. Because of their life-supporting nature, meals are important events when studying the individual's daily sequence of activities, since they must be allowed to break in between other activities. The time-geographic approach puts the full, unbroken activity sequence to the fore and it is based on the continuity of time in combination with the indivisible individual.

This relates to other important aspects of time used for activities, namely duration and occurrences. For example, each meal might take about 20 minutes to eat, and in the example above three meals occurred at different times in the whole activity sequence of the day, and other activities were performed between the meals. Consequently, the meals are seen as they appear in the context of other activities (contextual time use). If time is not regarded as continuous, but instead handled as a phenomenon that is divisible and additive (like money) it can be said that the person in the example eats for 60 minutes per day.[10] This is true in one respect, but from the perspective of an indivisible human being 60 minutes' eating per day is something very different, and less satisfying, than three occurrences of meals with a duration of 20 minutes each, interlaced with other activities in the course of the day, as shown in the sequential way of looking at time in a day perspective[11] (see Figure 2.3).

When regarding time as a continuous dimension, it flows along the Time axis, the duration of the activities in the sequence is revealed and thereby the uniqueness of every instant is underlined. This is close to people's experience from living their daily life. Contrary to this, when time is regarded as divisible and additive, sequence and duration are overlooked. Different conclusions might be drawn depending on the way in which time is regarded and what contexts the individual is involved in.

The time-geographic handling of the time dimension in an objectivistic way can be used for subjective purposes too. Using the objectivistic visualization of a person's daily activity sequence during a day (or other time perspective) as a

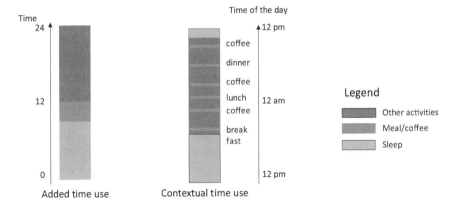

Figure 2.3 Visualization of added and contextual time use. An individual's use of the
24 hours of a day displayed as added time use (left) and contextual time use
(right). It is possible to account for the added sum of meals during a day from
a time-geographical illustration based on the continuous time. However, from
the added time use it is not possible to derive the number of meals, when they
occurred and their respective duration. In that case the only information is that
60 minutes per day are used for eating.

point of departure for discussion, people can make clear to others what they are
talking about. It helps people with different subjective impressions of the same
situation to communicate and talk about the context of when something happened.
In other words, there are gains on a subjective level from the objectivistic time
dimension. Examples of this will be presented in Chapter 7, regarding everyday
life applications.

About space and place

In contrast to time, which is immaterial and hard to capture, space and its places
make people experience distance and material restrictions on a daily basis. In
time-geography, space is the general concept for the geographical dimension,
while the place concept is used for specific locations in the geographical space. A
map shows the geographical locations of specific places according to agreed-upon
criteria for drawing maps; for example, the north–south and east–west directions,
legend and scale. When communicating about the location of places, knowledge
about such criteria is crucial. The space dimension used in time-geography is ideally based on a map constructed from agreed-upon conventions.

What is displayed on a map depends on the objects that are of special interest for specific studies, and various scales are used depending on the research
problem addressed. In the regional geography tradition the map is a central tool.

Locations of natural and human-made objects were mapped and regions were classified according to criteria. There was some kind of underlying belief that what was located at a place and what activities were performed at the time of mapping should continue. Hägerstrand opposed strongly to this way of reasoning, with its idiographic and descriptive approach. His goal was to discover general principles of change that all had a (temporary) geographical outcome. His studies on both migration and innovation were performed in the same region in Sweden, but his conclusions were general, not just valid for the locality where the empirical investigations were made.

Time-geographic studies most commonly concern the local and regional scales, but any other scale might be chosen, depending on the type of problem to be investigated (Hägerstrand 2004). On a map any material object with a spatial location might be illustrated. Time-geographers make far-reaching claims that researchers ought to consider the manifold types of individuals in several different populations existing in the area under study, but this must in some respect be overlooked for practical reasons since it is not possible to get everything into the same map. However, the mere awareness of this fact should help in considering the location of individuals in the most important populations in the region of interest.

Even if there are conventions about how to draw a map of a place, people might experience a place or a region in different ways. People are more familiar with what is located in the vicinity of places that they know well than with distant places. As experienced by subway travelers, the neighborhoods of the entrance and exit stations become familiar, while the geographical area on the surface between these stations may remain unknown territory.

The conventional map is a tool for communication about locations that is useful for people irrespective of where they are located within the area covered by the map. However, in the modern society where internet connections and tourist travel have exploded it might be argued that the situation has changed. However, in principle the situation is similar in the respect that the traveler will become more familiar with what is located and what places he becomes temporarily familiar with close to his hotel and other "new" places people experience (Shoval et al. 2015). People can now get to know places at a large distance from their home, which means that they know more places than people did before, but still there is a loss of detail in the knowledge as regards the territory between the places they visit (Cedering 2016).

The time-space and tools to think about it

The time-geographic notation system is an effort to provide a general tool for communication about phenomena not only as regards geographical location, but also the location in time. The notation system's main dimensions are time and space, as displayed in Figure 2.4. The time dimension shows the sequence of events as they appear before and after each other, while the space axis shows the location of the places where events take place, side by side.

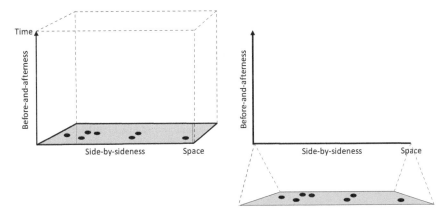

Figure 2.4 The basic dimensions in time-geographic visualizations: time, illustrating events in sequence, as they appear before or after each other; and space, illustrating the location of places, side by side of each other. Left: a traditional time-space cube, where individuals' movements are visualized in three dimensions. Right: a simplified visualization where individual movements are visualized in two dimensions.

About time-space

The roots of the efforts to combine the dimensions time and space can be found in the task Torsten Hägerstrand, as a PhD student in Geography at Lund University, was given by his professor.

Hägerstrand was asked to investigate what happened in a region in Sweden that had been left by many people during the large emigration from Sweden to America in the 19th century. By then the dominating theory said that the emigrants had left the houses that were abandoned in the landscape and Hägerstrand should study such abandoned houses from a geographic perspective (see the left part of Figure 2.5). By touring by bike in this landscape, searching for these houses, he found that they were located in areas with the worst opportunities for farming, with a lot of stones and low fertile soil. This location of abandoned houses was well in line with the theory of localization, indicating that poor soil quality yields poor harvests and therefore it is not attractive for people to live there. Hägerstrand and his fiancé Britt Lundberg (later his wife), who participated in the field work, began to study other dimensions than the geographic location, namely the living conditions of the people who had left the abandoned houses. They did it by studying the development of the population in the parish from the church registers wherein the priests noted births, deaths, people's ability to read and write, and migration to, from and within the parish. By looking deeper into what was registered as regards the living conditions of the people who had lived in the abandoned houses and how they had moved to, from and within the parish, Hägerstrand could show that the emigrants did not come from the abandoned houses. They were too poor to afford to buy tickets and go to America. Instead there was an opportunity for them to move into a house that was left by a less poor family who in turn had moved into a house left by a still less poor family who could afford the ticket to America by selling their house (see the right part of Figure 2.5). Of course, the abandoned houses were still located on the worst land, where it was difficult, if it was possible at all, to get enough harvest to feed households, so the location theory was valid, but the hypothesis that the emigrants came from these houses was not supported by the empirical results.

Through his studies of migration over time Hägerstrand identified migration chains, which reveal which individual in a population moved where and when in a time period (Hägerstrand 1957). Hence, he followed individuals in time and space. This was something new in geographical research at that time. However, today this is common knowledge in geography.

Without the use of a variety of sources dismantling the processual character of people's dwelling history, it had not been possible to discover the migration chains. One prerequisite was that each individual was regarded as an indivisible whole. Hägerstrand's studies underlined the need both for a map showing the location of houses and the register books revealing the time when a particular individual moved from one place to another. Thus the seeds for combining the space and time dimensions in a dynamic approach were planted in Hägerstrand's mind and, according to what he said at a time-geographic seminar in Lund in the late 1990s, formed the foundation for the ideas that were later developed into the time-geographical approach. When Hägerstrand formulated the time-geographic approach in the 1960s, the concept of the individual path (see Chapter 3) was derived from this kind of individual movement from place to place.

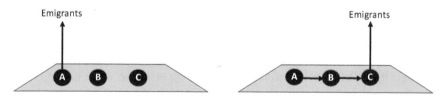

A = abandoned house, B and C inhabited houses

Figure 2.5 Two different ideas about abandoned houses and emigration from Sweden. The left part of the figure shows the dominating hypothesis about emigration and abandoned houses: A is an abandoned house and the family who left it were supposed to be emigrants to America. House A is located in an area with poor farming conditions. The right part of the figure shows that the inhabitants of the abandoned house had moved to house B and the former inhabitants of house B have moved to house C, from which a family that was bit more wealthy had moved and emigrated to America. A chain of movements was started by the emigration of the wealthier families (even if they were not rich).

The migration chain concept was one of Hägerstrand's innovative results from the research in Asby parish, which showed that places (locations of houses with their inhabitants), time (as a continuous process) and the movements of individuals between different places, taken together had vital importance for understanding the development of the region (Hägerstrand 1950; 1957).

A wider theoretical conclusion from this research concerned the importance of investigating what happens in a region continuously over time and not just looking at single points in time. Hägerstrand argued for considering time as a continuous dimension to show the changed location of people in a region during a period of time. In Figure 2.6 the difference between time as a continuous process and time as points in time is made clear. Without a continuous time dimension it is not possible to see who has stayed at a location, who has moved where from and where to in the region. By following the path in time and space, the basal thinking behind Hägerstrand's migration chains is made obvious.

In the upper part of Figure 2.6 the localization of the individuals in the population is seen from the point-in-time perspective. The population is not evenly distributed over the region at any of the time points. Since the dot patterns differ between the points in time, it is obvious that there has been some change in the location of the individuals between the time points (t_0 and t_2). One place has grown in terms of number of individuals (e.g. the concentration of dots in the upper part of the region at time point t_2) while other places have lost most of their inhabitants (compare, for example, the few dots to the left in the lower part of the

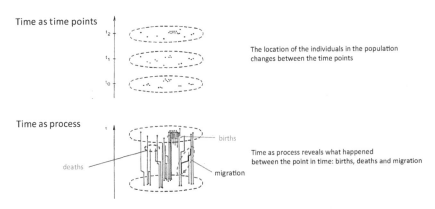

Figure 2.6 Time as point in time and time as process – fundamental differences. Looking at time as time points (the upper part of the graphic) gives no indication of who has moved and who has stayed. Information is given that some people live at some places. Time as process (the lower part) reveals not only who has moved and where to, but also other population changes, like births and deaths. Based on a figure in Hägerstrand 1993.

region at time point t_2 to the dots at time point t_0). Each of the three points in time provides a snapshot of the localization of the population. However, what has happened between them remains hidden.

The lower part of the figure represents exactly the same region, with exactly the same population located at the same places during the same time period as in the upper part of the figure. The difference is that the lower part, by showing time as a continuous process, also reveals what happened during the time period between the points in time in terms of how each individual in the population has moved or stayed. By drawing a line to indicate the movements of each single individual, each person's movements in the region can be followed, and here the basis of the time-geographical individual path concept is identified. Some individuals stay at the same place over the whole period and they are illustrated only by vertical lines. Some people die, indicated by the line ceasing to exist; others are born, indicated by new lines starting during the period, which happens towards the end of the period and only in the location where the population grows. Finally, some lines illustrate the movements of individuals from one place to another within the region – the line is angled between its vertical sections. Some lines describe how a person moves once, and there is an example of a person who moves twice time during the time period. This is the base of the notation system that is thoroughly considered in Chapter 3.

From the discussion above it is clear that the existence of material phenomena can be described by their place location during a time period. Some material phenomena move by themselves (humans and animals), while others are moved by external forces (human or natural) from place to place – in both cases, however, the material phenomenon occupies both time and a place, thereby excluding other material phenomena from being at exactly that place during exactly the same time period. Concepts are needed that put to the fore the material existence of phenomena at various places over time, concepts which can be used to describe the couplings of events in both time and space. Such concepts increase the ability to conceptualize, describe, analyze and understand processes wherein individuals are involved and are located at different places over time. One main effort of the time-geographical approach is to contribute perspectives on processes that can help researchers to avoid "escap[ing] from the doctrine of perpetual flux", to borrow the words of Russell (1961), quoted above.

In one of his last articles Hägerstrand wrote:

> The main purpose of the approach which came to be called time-geography was to open up a related perspective from the outside in which the main issue is how the myriad of objects in our life-world, i.e. all that existing upon the earth's surface, get placed or place themselves through contact with one another during the lapse of time.

> (2004: 323)

Notes

1 Space is used as a general concept, while place is used for specific locations in this space.
2 Kant gave lectures that divided geography from history, but he also claimed that the two are closely related. There has been a huge debate about this, summarized by Louden (2015), where May (1970: 121) is cited, saying that "Kant exhibits considerable ambivalence at times respecting the issue of clearly separating the history of nature from the 'description of nature', or geography". Also, Harvey (2000: 554) is cited, claiming that "the Kantian prescription to construe geographical knowledge as mere spatial ordering, kept apart from the narratives of history".
3 This tradition in geography has been more or less abandoned since the late 1960s but at the time when Hägerstrand was studying and in the first part of this career it was a dominating direction in geography. Looking back at his time as a geography student, Hägerstrand (1983: 244) wrote: "Lectures in regional geography were abominably boing . . . Geography appeared not as a realm of its ideas or a perspective in the world, but as an endless array of encyclopaedic data."
4 The Nobel Museum in Stockholm developed a physical model of the individual paths in the time-space (visualized by silver threads over a map of the world) of laureates in economics. The importance of Chicago as an inspirational point is obvious.
5 At least at the level of human life.
6 In Chapter 7 three different perspectives on time are presented and discussed (daily life perspective, instrumental perspective and constructive perspective).
7 This will be shown in the chapter about applications of the time-geographic approach, for example when it is used as a tool in occupational theory practice and research.
8 It is very hard to visualize movement in a static figure!
9 Also, memories change over time and other perspectives might open.
10 Each meal endures for about 20 minutes, and $20 \times 3 = 60$.
11 The time-geographic way of looking at time as a continuous dimension implies that the information about (1) the three eating occurrences, and (2) the duration being 20 minutes per meal, is important, and this is fundamentally different from saying that the person eats for 60 minutes during a day.

References

Cedering, M. 2016. Konsekvenser av skolnedläggningar. En studie av barns och barnfamiljers vardagsliv i samband med skolnedläggningar i Ydre kommun. *Geographica* 8. Uppsala University, Department of Social and Economic Geography. Diss.

Hägerstrand, T. 1950. Torp och backstugor i 1800-talets Asby. In *Från Sommabygd till Vätterstrand*. E. Hedkvist (ed.), pp. 30–38. Linköping, Sweden: Tranås Hembygdsgille.

Hägerstrand, T. 1957. Migration and area. Survey of a sample of Swedish migration fields and hypothetical considerations on their genesis. In *Migration in Sweden: A Symposium*. D. Hannerberg, T. Hägerstrand and B. Odeving (eds), pp. 27–158. Lund Studies in Geography, Series B Human Geography, No. 13. Lund, Sweden: C.W.K. Gleerup.

Hägerstrand, T. 1970. What about people in regional science? *Regional Science Association Papers*, Vol. XXIV, pp. 7–21.

Hägerstrand, T. 1972. Om en konsistent individorienterad samhällsbeskrivning för framtidsstudiebruk. *Ds Ju* (1972): p. 25.

Hägerstrand, T. 1983. In search for the sources of concepts. *The Practice of Geography*. A. Buttimer (ed.). Harlow: Longman Higher Education, pp. 238–256.

Hägerstrand, T. 1985. Time-geography: focus on the corporeality of man, society, and environment. In *The Science and Praxis of Complexity*, pp. 193–216. Tokyo, Japan: United Nations University.

Hägerstrand, T. 1993. Samhälle och natur. Lunds universitets geografiska institutioner. *Rapporter och Notiser*, No. 110, p. 21.

Hägerstrand, T. 2004. The two vistas. *Geografiska Annaler*. Series B Human Geography, Vol. 86, No. 4, pp. 315–323.

Louden, B.R. 2015. The last frontier: exploring Kant's geography. In *Reading Kant's Lectures*. R.R. Clewis (ed.). Berlin, Germany: Walther de Gruyter, pp. 505–525.

Russell, B. 1961. *The history of Western philosophy*. New York, NY, and London: Simon & Schuster/George Allen & Unwin.

Shoval, N., McKercher, B., Birenboim, A. and Ng, E. 2015. The application of a sequence alignment method to the creation of typologies of tourist activity in time and space. *Environment and Planning B: Planning and Design*, Vol. 42, No. 1, pp. 76–94.

3 Time-geographic concepts and notation

Food for thought

Hägerstrand experienced the fundamental transition of the Swedish society in the mid-20th century and this strongly influenced his worldview. There were improvements of people's living conditions and material standards and these new conditions, of course, influenced experiences, values and ideas about what it is possible to do in the society (see Chapter 1). Many restrictions once set up by the vertically linked, short-distance society were eased, thereby increasing people's opportunities to get education, move from rural areas to the urbanizing cities, change jobs, meet new people and move up the social ladder on a never-previously-seen scale. There were changes in many aspects of life; materially, regulatory and culturally – but at the same time new restrictions emerged.

Hägerstrand found food for thought in a chapter by Karl Popper in a book edited by Popper and fellow philosopher John Eccles (1977). Karl Popper elaborates on various aspects of human existence. Hägerstrand wrote about Popper's ideas in a thought-provoking article (1985) referring to these ideas about three interrelated aspects, or worlds, of human existence. World 1 is the material world, with all its existents, living and non-living, occupying space and taking place with their bodies. World 2 is the world of thoughts, ideas and phenomena that dwell in people's minds and brains. World 3, finally, is the world of cultural constructions, rules and formulations. In real life, the three worlds are complementary and interlaced and hard to distinguish from each other. For example, even though thoughts belong to World 2, thinking is hosted in the brain, which belongs to World 1, the material world. New expressions of culture, laws, rules and conventions of World 3 develop over time in a steady interplay between the material world (World 1), the ideas and thoughts of humans (World 2) and the previously developed cultural expressions of societies (World 3).

Hägerstrand's time-geography, as indicated in Chapters 1 and 2, is materialistic in the sense that the bodily existence of material phenomena (World 1) in the time-space is regarded as a fundamental point of departure. Time-geographic descriptions, visualizations and analyses of processes build on the continuous time-space movements of material individuals between their birth (or creation) and death (or destruction). Hägerstrand was criticized for being materialistic and not taking people's thoughts, emotions, experiences and feelings into account in his analyses. Hägerstrand never denied the importance of the immaterial dimensions of existence.[1] However, he was concerned about social scientists underemphasizing the material dimension of existence. His opinion was that social science should not leave the material world over to natural science and technology, since materiality is fundamental for many aspects of living life, not least the social and emotional.

Music, belonging to Popper's World 3, was an important source of inspiration for Hägerstrand and his development of the time-geographic notation system. Music is instant; tones are sequentially played and the sound disappears as soon as the tones are played, while the memory of the previous tone and the current hearing of the next, and the next, and so on, altogether constitute the impression of the melody, a sequence of tones. Hägerstrand found similarities to the spoken language, and with people's activities during over time. Words disappear once they are pronounced and one activity is replaced by another and yet another, and both language and activities are sequentially performed. His notation system is a way to replicate the musical score, with its signs for how to play the instruments, but instead of a description of a melody with tones, tempo and strength, the time-geographic notation system is a description that shows the sequence of activities performed by an individual, including movement activities, in the time-space. The result might be satisfactory, positive or negative for the individual, and it might fall out according to the plans set up, or not.

The notation system has been developed to investigate how and why individuals, living and non-living, come together (couple) and leave each other (de-couple) in various ways in the time-space. To analyze the complexity in the emergence of such couplings and de-couplings there is a need for concepts defining what kind of event is going on and what individuals are included. In this chapter, time-geography's concepts and its notation system for handling this challenge are presented.

All concepts in the time-geographical approach are imbued by the idea of time as a continuous dimension with its main constituents: past, now and future. *Individual, path, elementary event, bundle, prism* and *population* are the basic and most general time-geographical concepts useful for describing, analyzing and communicating about processes that relate to phenomena with a material existence in the time-space.

In addition to these basic concepts, there are three time-geographical concepts that are especially suitable for investigating the complexity of human activities, *project,*[2] *constraint* and *pocket of local order.*[3]

Basic time-geographical concepts

Individual

In most social science, the individual[4] concept is used for human individuals, people, but the time-geographic approach widens the denotation of the concept. Hägerstrand was inspired by abstractions and general concepts and claimed that among the self-evident aspects in the world is its grain-structure, ranging from micro to macro scale (Hägerstrand 1985: 196). The 'grains' exist continuously over some period of time. Grains with unbroken existence (from birth to death as regards living entities and from construction to destruction as regards non-living entities) are indivisible at a chosen scale and the time-geographic concept chosen for them is individual.

Consequently, in the time-geographic approach, the individual concept is used for a human individual as well as for an animal, a plant, an artifact or a material thing.[5] Thereby the concept is an abstraction for all existents (or 'grains') in the time-space. Since each individual is indivisible it cannot be divided or truncated without losing its meaning as an individual. As already mentioned, a living individual has a limited lifetime, starting with birth and ending with death. However, after death an individual that once was living is divisible; it dissolves and its constituents are integrated into other individuals (e.g. worms, soil, plants). Non-living individual things, artifacts and creatures of nature are regarded as indivisible at a chosen level of investigation from their creation until destruction or taken apart.

Then, what is regarded as an individual at one level may, at a more detailed level, be regarded as composed of several individuals of different kinds. For example, a car is an individual at one level where it has specific capacities. These capacities get lost if the car is taken apart. However, the parts or components that together make up a whole car are themselves individuals of different kinds at a lower level, like an engine individual, a steering wheel individual, and so on. Taken one by one these lower-level individuals of different kinds cannot produce the capacities that characterize the higher-level individual car as a whole indivisible individual. The components of the whole car must be coupled to each other in a certain way to gain the capacities of the car.

A human, regarded as an individual in the time-geographic sense, has capacities that other kinds of individuals lack; for example, spoken and written language, intentions, strategic thinking, capacities to formulate ideas, imaginaries and plans for the future. Thereby, humanistic and social perspectives on human individuals' actions are crucial. However, at the same time, and just like the bodies of other kinds of individuals, the individual human being has a material body that exists in time and is located in space.

The individual concept is a basic part of the time-geographic approach and it is vital to specify what kind(s) of individuals are in focus when doing empirical studies.

Path

The path[6] concept is introduced to visualize and follow the time-space movements of individuals. The movements of an individual always follow along the time and place dimensions and the path is constantly created as the now transforms future into past. However often foreseen, one core point is that even when the individual stays at the same place there is a movement in time. The path, then, is a visual means to increase and widen the comprehension of the processual thinking in time-geography and it is a foundational and fundamental concept in time-geographic thinking and the notation system. The combination of the individual and path concepts into *individual path* reveals that the path illustrates the time-space movements of an individual. Hence, it can be used to visualize the movements of all different kinds of individuals (living and non-living) and it works on various scales in time and space.

An individual path visualizes movements that have already happened, and the path is eventually created as now transforms future into past. Figure 3.1 shows the creation of an individual path by illustrating the movements of a human individual during a short time period: Mr. Svensson is going to the grocery to buy food ingredients for a party in the evening.

The upper-left part of Figure 3.1 shows a map, and the spatial movements of Mr. Svensson are indicated with a line along the road. However, such a static picture reveals little about where he is located over time. The rest of the subfigures in Figure 3.1 show the consequent creation of the individual path illustrating the time-space movements of Mr. Svensson, where the Time axis (y-axis) shows the before-and-after relations and the Place axis (x-axis) shows places (side-by-side-ness) in the geographical space. The upper-middle part of Figure 3.1, then, shows the time-space foundation where the individual path following Mr. Svenson's time-space movements is illustrated: initially he is alone in his home, planning his shopping, which takes time, and consequently the individual path illustrating his activities moves parallel to the Time axis, while located at the same spot on the map. This indicates that he moves in time (the now is moving along the Time axis) but he is at the same place, stationary in space (in his home). When he has made up his list for shopping he goes to the shop (upper-right part of Figure 3.1) and then the individual path illustrating his time-space movements turns away from his home towards the grocery, and now illustrating movements both in time (along the Time axis) and in space (along the Place axis as he moves between places). When he enters the shop, he does his shopping and then he is again geographically stationary and he moves only in time (lower-left part of Figure 3.1). This is of course an oversimplification since he moves also within the shop, but at this scale (where the shop is regarded as a location) the individual path is, again, parallel to the Time axis since he is located in the shop for a time period. When he has paid he goes homewards and the individual path reveals that it takes more time to go home than it took to go to the shop, since he has heavy bags to carry (middle part of Figure 3.1). The slope of the individual path, then, reveals that he moves at a lower speed while walking home than when he walked to the shop. Back home he unpacks the goods (shown in the lower-right part of Figure 3.1) and the individual path visualizing Mr Svenson once again is moving parallel to the Time axis, but it is stationary on the Place axis (in the home).[7]

It is important to underline that the path is an abstract illustration of the time-space movements of an individual, while the individual represented by the path is something much more complex. Consequently, an individual path increases the knowledge about time-space movements of the material body of humans as well as of other kinds of individuals. It puts to the fore the often-neglected fact that time also passes when an individual is stationary.

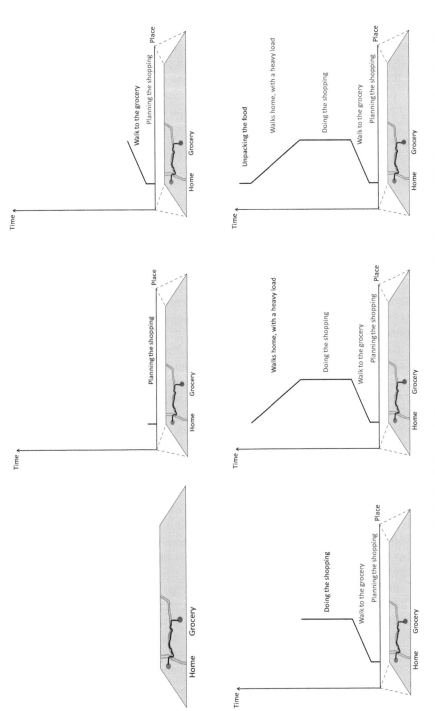

Figure 3.1 The construction of an individual path. The individual path is demonstrated with Mr. Svenson and his shopping for ingredients for a party he will arrange in the evening.

All individuals' time-space movements are created in the now and each individual path illustrates an individual's approaching, staying at and leaving places. The path can be regarded as a flow of events, experienced by a human individual as her activity sequence. In addition, the path can indicate the individual's getting in touch with other individuals, staying at a place doing something together with them, and leaving it afterwards. The paths of two individuals show their time-space couplings to each other, and when coupling they form a bundle. For closer analysis of such time-space movements and of couplings and de-couplings, as these are illustrated by individual paths, the time-geographic concept of the elementary event has been created.

Elementary event

The elementary event concept is sparsely promoted in time-geographic literature. However, in Hägerstrand (1970: 14) the concept "large series of small events" is already used and it may be the seed for what grew into the elementary event concept in Hägerstrand 1974 and 1985. Hägerstrand writes: "Life becomes an astronomically large series of small events, most of which are routine and some of which represent very critical gates" (1970: 14). Some years later he briefly introduced and discussed elementary event as a time-geographic concept (Hägerstrand 1974; 1985), but he did not go deeper into it until in his final book (Hägerstrand 2009: 104–113).

Each individual exists for a period of time,[8] which for living individuals is limited by birth and death and for artifacts and natural things by creation and destruction. During this lifetime, living individuals need resources to produce and use food and shelter, and in addition they need rest, care and social relations. To satisfy these needs they depend upon other individuals, both members of their own population (e.g. older generations, children, partners) and individuals from other kinds of populations (food, buildings, tools, and so on). This means that living individuals must be able to approach and get in touch with both living and non-living individuals from various populations. In his final book Hägerstrand (2009: 76) wrote that the most fundamental of all events is encounter. When an individual cannot encounter another individual of importance, the latter individual is out of reach for the former. If food individuals are out of reach for a living individual, there is a risk of hunger, or even starvation. In order to analyze what human individuals do to satisfy their needs it is important to use concepts that specify how people navigate between locations where they can reach the necessary other kinds of individuals (from different populations) over time.

The concept for specifying such navigation in the time-space is elementary event and various elementary events denote different specific parts of such a navigation process. There are elementary events designed for describing the movements of single individual paths and elementary events for specifying the time-space couplings of two of more individual paths. Here, the elementary events for single individual paths are presented first, followed by elementary events for multiple paths.

Elementary events describing the time-space movements of a single individual path are: *move, arrive, stay,* and *leave* (see the left part of Figure 3.2). Of these events, arrive and leave happen in an instant, while move and stay usually have a longer duration. The elementary event move has two different but related meanings. First, it concerns *move to* (encounter) or *approach* a place, and second, it is about *moving away from* or *receding from* a place, both thereby indicating geographical moves.[9] However, differently from most other approaches in social science, in time-geography stays are regarded as a kind of movement, but during a stay the individual moves solely along the time dimension since the geographic location does not change.

The left part of Figure 3.2 shows some elementary events of the individual path of Mr. Svenson, who goes from his home to the grocery shop and back home again before arranging his party (compare it to Figure 3.1). Mr. Svenson cannot do his purchase if the shop is not open and if a shop assistant is not there, and in the right part of Figure 3.2 there are two individual paths, one for Mr. Svensson and one for the shop assistant. The individual path of the shop assistant can be described by one single elementary event for the whole period: stay, since he is in the shop ready to assist any customer, both before and after he has served Mr. Svensson. When Mr. Svensson has arrived at the shop, he gets in touch with the shop assistant and they stay in touch until Mr. Svensson leaves in order to move homewards. What is communicated between the two during the Mr. Svenson's stay in the shop is not revealed by the visualization.

The elementary events specifying the time-space couplings between two or more individuals that relate to each other in the shop are, for example, get in touch, stay together, and disengage (see the right part of Figure 3.2). *Get in touch* means the very moment when two (or more) individuals of any kind come together at a place at a time. Before getting in touch at least one of them must have moved to (encountered) the location of the other(s) and after getting in touch they may stay together for a period of time. Some time later one (or more) individual(s) may find that it is the right moment to disengage and leave, and after leaving this individual moves from the other individual(s).

An example from the industrial production of automobiles might illustrate different levels of elementary events where human individuals relate to material things over time. Humans construct cars and people can utilize the specific capacities of the car to *move from* one place and *to* another. For example, an employee in an automobile factory uses a car to move to her workplace on working days. She *stays* there to do the job until the work day is over, and she *leaves* the workplace and uses the car to move homewards. At this level the car is a tool for the human individuals to *approach* and *depart* from the workplace and the home, respectively.

On another level, a car in production is assembled from many lower-level individual parts and components. In the production workshop each car is assembled as result of humans and machines making the many different kinds of individuals (parts and components) move to (encounter), arrive, get in touch and stay together. The meaning of the assembly is to make the components and parts stay together and eventually a car with its specific capacities is completed.

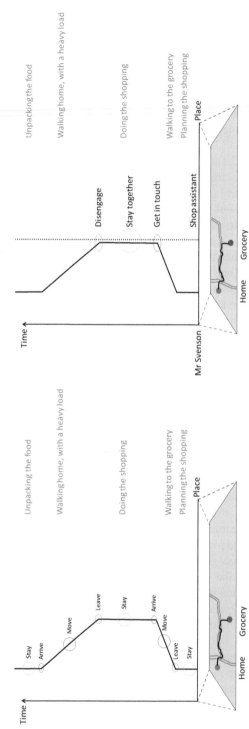

Figure 3.2 Elementary events. Left part: elementary events of a single individual path, exemplified with Mr. Svenson going shopping. The elementary events displayed are: *move; arrive; stay; leave.* Right part: elementary events describing the individual paths of Mr. Svenson and the shop assistant. The shop assistant is stationary all of the time and there is just one elementary event for his geographical position: *stay.* The individual paths illustrate the complementary elementary events for the coupling between Mr. Svenson and the shop assistant during the shopping activity (*get in touch; stay together; disengage*).

There are other elementary events, which concern individuals that lose their identity when another individual exerts power and ends the existence of the former as an individual. It might happen when human individuals satisfy their need for food and utilize other kinds of individuals (like vegetables, meat and other sorts of food) for preparing a meal, and eat it. The elementary events are, for example, *intake, mold/digest*, and *fuse/extract*. In terms of elementary events, the food ingredients are moved and they get in touch with each other via the hands of the person cooking, and then they stay together as a meal until eaten by the human individual. When the meal is digested and transformed into energy in the human body, the remains leave the body as faeces and urine.

When an individual's existence in an area is regarded over a period of time, the elementary events can be used to identify its movements and stays, and to make clear this specific individual's time-space relation to other individuals in the area. It is a way of communicating the time-space process of individuals in different constellations. There is a lot of information given even by one individual path with its various elementary events. A complex pattern is shown when more individuals of one or more different kinds are included, forming a bundle.[10]

At an abstract level, then, the elementary events concept helps specify the precise time-space movements of one or more individual(s) as described by the individual path(s). The abstract use of the individual concept makes it possible to consider the time-space interrelation between individuals of different kinds during a period in a region.

Bundle

Hägerstrand suggests that the bundle concept refers "to a grouping of several (individual) paths" (1970: 14). The concept describes the staying together of two or more individuals at the same place or on the move (for example, in a bus or on a walk) during the same period of time. For example, within a home and taken together, individual paths visualizing the members of a household, their possessions and materials can be described as a bundle, while in a factory a bundle describes the time-space arrangement of employees, materials and machines. A bundle, then, is visualized by individual paths (see Figure 3.3), and the elementary events show when the individuals involved move to, arrive, get in touch, stay, leave and recede or move from a place. Hence, they reveal the sequential order of the individuals' respective arrival at and leaving of the place and inform about the time that the involved individuals spend together at the place.

Prism

The individual path illustrates the time-space movements of an individual in the past and it might serve as a kind of protocol up to now. The path cannot cross the now-line. Consequently, the individual path continues until the now, and as the now moves on (transforming future into past), the path is constantly created and illustrates the individual's time-space movements. The path, then, has to do with what has happened. Existence, however, also concerns what might happen in the future.

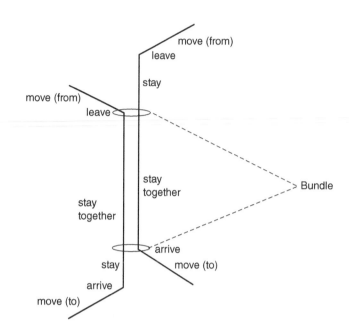

Figure 3.3 A bundle with two individual paths, also including stays and movements of the two individuals illustrated by the paths.

The opportunities to perform activities in the future are important for living creatures, not least for human individuals. They make up plans and have expectations for what to do, and estimate what claims these plans have regarding the future location of the individual herself in the time-space.

The time-geographical concept for capturing the future possible locations in the time-space of an individual is *prism*.[11] At each now, the individual has a specific location in the time-space and from this now the individual has opportunities to depart and approach other places in the future (see the upper-left part of Figure 3.4), which displays the possible moves away from a location 'now' at a given speed. In the upper-right part of Figure 3.4 a restriction is included, namely that the individual has to be back at the place of origin at a certain point in time, and this makes up the prism. The shape of the prism depends on (1) the geographical location of the individual at the point of departure in time (now); (2) when in the future the individual has either to be back at the same place, or to be located at another place; and (3) the speed of the means of transportation available to the individual.

Irrespective of how many opportunities there are within the prism of an individual, just one of them can be realized each moment as the now moves along the time axis. This follows from the assumption that the individual is indivisible and as such it can be located at only one geographical location at every moment. The prism shrinks as the now moves along the Time axis if the point in time when the individual has to be geographically located in the future stays the same as it was

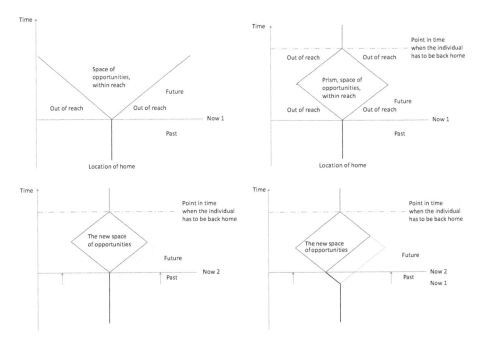

Figure 3.4 The principle of the prism. Upper-left part: the movement possibilities for an individual with no restrictions regarding future location. Right: a prism, the future space of opportunities for a person being at home and who has to be back home at a given point in time in the future. Lower-left part: the prism shrinks as the now transforms future into past (from Now 1 to Now 2). Right: the shape of the prism changes if the person starts moving from home at Now 1.

at an earlier now, and if there is no change in the speed of transportation means (the lower-left part of Figure 3.4). Thereby there are fewer opportunities within the prism at the later now than at the former. The shape of the prism changes if the individual has moved to another place (the lower-right part of Figure 3.4).

Lenntorp developed the prism mathematically and empirically in his PhD thesis (1976). He discussed different shapes of the prism depending on various means of transportation. He also discussed what errands people with various transportation opportunities can perform within the limits of their specific prism volume in the future. For this purpose, he put together activity sequences ("daily programs") and tried out which of them that could be fulfilled given various means of transportation and timetables in public transport.

Hägerstrand meant that humans easily refrain from taking the materiality of existence into consideration when planning for the future. Thinking encouraged by the prism may assist in considering the material world (Popper's World 1) when creating plans (Popper's World 2) for the future: "It is as if our well-developed capacity to store and hold together systems of ideas makes us unable intuitively to feel the limitations of the external world to accommodate our projects" (Hägerstrand 1985: 214). However, not just the limits of the prism influence what projects are possible to realize in the future. There are other constraints of importance on what it is possible to do, and these will be presented later in this chapter.

Population

A population[12] consists of individuals of the same kind, which over a period of time exist in a defined area. The time-geographical use of the *population* concept denotes all individuals of the same kind existing within a defined region during a time period. In any region, there are many populations: plants, animals, human beings, artifacts, roads, and so on. In the landscape the varieties of individuals in different populations serve as prerequisites for the survival of some individuals and death of others. For example, on a pasture the individuals in the cow population eat straws in the grass population, thereby limiting the ability of the grass to grow a bigger population.

When considering the historical background for human individuals' life histories the population concept and the variety of the individuals within it is of interest. For example, admission to higher education and the right to vote have different meanings to people of different generations but in the same population. A previous decision, which currently concerns all individuals in a population, has a different influence on people depending on when they were born in relation to the time of the decision. The oldest individuals in a population might not have been able to enjoy the rights because the time in their life when it would have been of importance has already passed. In Sweden, for example, women were allowed to vote for the first time in the 1921 general election, and women were not allowed to study in the public high school system until 1927. These two decisions obviously formed the life opportunities for many generations of women. Women who had passed the age of going to high school when the law changed could not get that level of education (see Figure 3.5). This resulted in them being excluded from many jobs in the labor market where a high school exam was a condition. This had long-term consequences for the development of equal opportunities of men and women.

Individual and aggregations of individuals

With inspiration from his studies on migration Hägerstrand suggested that necessary links exist "between the micro-situation of the individual and the large scale aggregate outcome" and that these must be recognized (Hägerstrand 1970: 9). His idea was to avoid handling people in the same way as money or goods, as divisible entities, and instead make it possible to identify the unique individual as a whole also after aggregation. To underline what he meant he quoted a student's saying: "We regard the population as made up by 'dividuals' instead of 'individuals'" (Hägerstrand 1970: 9).

In a human population there are some resemblances between the individual members, which stem from the human need for sleep and food; compare the

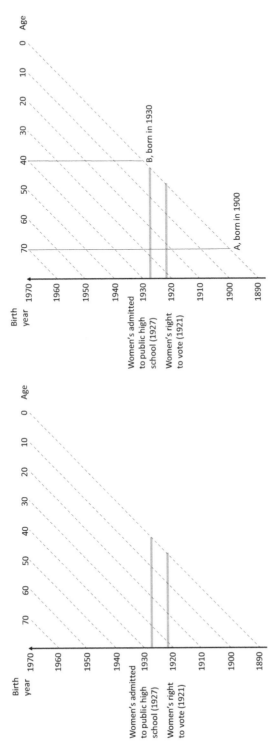

Figure 3.5 Formation of generational experiences from societal reforms. Left: two important reforms for women in Sweden – the first general election where women could vote (1921) and the admission of women to public high schools (1927). Right: the different opportunities for women – A, born in 1900, and B, born in 1930 – to gain from the reforms. Woman A could vote in the 1921 election but she could never enter the public high school since she was too old for school in 1927. Women B could both vote and go to public high school.

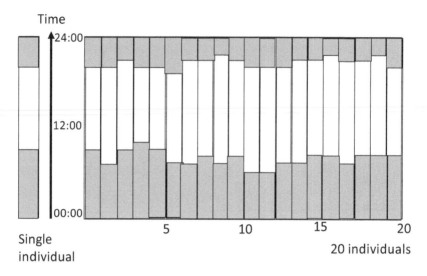

Figure 3.6 Contextual time use at individual and aggregate levels. Contextual time use
for one individual (left) and an aggregate of 20 individuals (right).

contextual way to visualize time use at the individual level in Figure 2.3. An
illustration of an aggregate of human individuals (a population) shows that there
are some variances in sleeping patterns, but that there is still a common everyday
rhythm. Figure 3.6 illustrates simplified contextual time use (limited to sleep and
other activities) for one single individual and for a group of 20 individuals, which
indicates that there is a pattern at the aggregate level building on the similar, but
not identical, daily contextual time use of many individuals.

Complex time-geographic concepts related to activities of living individuals

In time-geography, there are some more complex concepts that are specifically
relevant for living individuals and their biological or genetic programs and ability
to set up goals for activities. These concepts are *project, program, constraint* and
pocket of local order.

Project and program

The time-geographic use of the project concept relates to human individuals,
while *program* relates to human beings as biological creatures and to other animals. Here, the project concept in its time-geographic sense is presented.

The word *project* comes from the Latin words *pro* (forth) and *jacta* (to throw)
and means that something is thrown forth. A project, then, relates to an idea about
something or a state in the future. A project is created when one or more person(s)
set up a goal for the future, produce a plan for activities and make efforts to follow

the plan in order to achieve this goal. For example, things produced by humans are the results of people's realization of previous projects as plans for achieving goals based on what, by then, were ideas about the future.

In time-geography, the project concept is suggested to "designate the entire set of space/time-uses of people, things, and room leading up to some goal" (Hägerstrand 1985: 201).

A goal or an idea is created and relates to something that is wished for to be realized in the future; it might be a material or immaterial result. For example, an individual might initiate a project to knit a sweater. This project's direct goal is to have a sweater in the drawer for future cold days. However, the knitting-of-a-sweater project might also be part of a larger project, like "wellbeing", where knitting is a recreational activity. This means a project may have different kinds of goals and they may be integrated into longer-term projects. The realization of the sweater project is the performance of knitting activities going on as the now-line moves and transforms future into past. Of course, there are not many people who knit a sweater without pauses for meals, sleep, work and other activities, so the project is split up and its activities pop up now and then between other activities in different projects during a longer time period. The project's direct goal is achieved when the sweater is ready and exists in the material world as an individual artifact of its own. The sweater is thereafter ready to use by people who have it within reach. The product consists of embedded time, material resources, knowledge and skills.

Projects and their goals are based on ideas and since ideas are immaterial, it is not possible to visualize them in the same way as material phenomena. In an article from 1985, Hägerstrand presented a way to widen the time-geographic visualization in order to consider people's ideas, thoughts and wishes in the visualizations. This relates to the philosophically based view of Popper (1977) presented in the beginning of this chapter. Hägerstrand showed in his article that even though time-geography has its anchoring in the material world (World 1), this world is closely related to the subjective World 2, with human minds and mental states, and also that the cultural products of World 3 are important for the outcome. Hägerstrand writes, "Note that a book containing, say, a poem, belongs to World 1, whereas the poem as such does not. It is a World 3 entity" (Hägerstrand 1985: 194).

Similar thinking is behind the time-geographical visualization presented in Figure 3.7, based on a human individual's perspective.[13] This individual's existence is described from two positions: the Outer world and the Inner world. The Outer world, displayed to the right in the figure, shows what has happened in the past, the current situation and what future opportunities there are. The Outer world can be visualized with the time-geographical concepts presented above: the individual path (what has happened in the past) and the prism (opportunities in the future given the principle of return and the location now in the time-space). The Inner world of the person is displayed to the left in the figure, and includes her experiences, intentions, wishes and wants. In the Inner world of a person, now is probably not just the moment where future is transformed into past (the now-line), but instead now is more of a period, or a "now-zone". The Outer world resembles Popper's World 1 and the Inner resembles Popper's World 2.

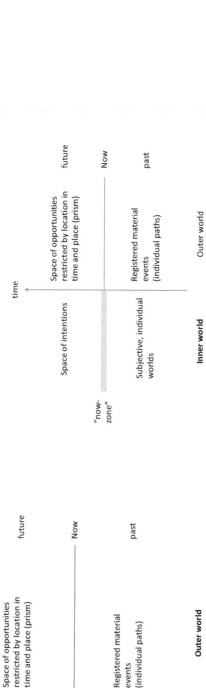

Figure 3.7 The Inner and Outer worlds of an individual from a time-geographic perspective. Left: the Outer world (like Popper's World 1), as illustrated from a time-geographic perspective, with now as a distinct partition between past and future. Right: the Outer world relates to the Inner world (like Popper's World 2). The past in the Inner world hosts the subjective dimensions and experiences of the individual. The future in the Inner world relates to the individual's intentions. There is no distinct now in the Inner world; it is more like a now-zone.

Figure 3.8 illustrates the continuation of the story of Mr. Svenson and his project to arrange a party at home in the evening (compare it to Figure 3.1). His shopping for groceries is considered in the context of his Outer and Inner world experiences. The left part of Figure 3.8 displays the individual paths of Mr. Svenson and the shop assistant in the Outer world. There is also an additional individual path, symbolizing a shop assistant working in another grocery located nearby. While the individual paths of Mr. Svenson and the shop assistant stay together during Mr Svenson's shopping, something happened that caused Mr. Svenson to start reconsidering his shopping in that specific shop: the shop did not supply ecological meat. This is indicated with a dotted line from the Outer to the Inner worlds of Mr. Svenson in the right part of Figure 3.8. Mr. Svenson cares a great deal for animals' health and wellbeing, and he became very angry (which is implicated in the Inner world in Figure 3.8) and he immediately thinks for himself that he will never go back to this shop again. This, then, is his intention for the future, as indicated in the Inner world in the figure.

However, as his cooking for the evening party proceeds he suddenly discovers that a key ingredient is lacking (see Figure 3.9). He has to go and buy it, and now his intention to not go to the shop where they do not supply ecological meat is challenged. As his prism in the upper-right part of Figure 3.9 shows, he does not have enough time to go to the alternative shop. Instead, he has to rush back to the shop he did not want to go to and buy the ingredient – if not, he would not be home when his guests will arrive. The clash between the idea (Popper's World 2) and the materiality (Popper's World 1) is revealed by the specific time-space location in the time-geographic visualization.

This is a simplified example of how time-geography facilitates in understanding the problem that arises when intentions and goals clash with what it is possible to realize in the material world, and, consequently, what comes out of them. Hägerstrand wrote about this in his early publications, and it is related to the time-geographic coupling constraints presented below (Hägerstrand 1970; 1976; 1985).

Projects can be individual, as exemplified by Mr. Svenson and his party project above, or organizational. Individual projects relate to what goals an individual wants to achieve for herself from her activities, while organizational projects relate to goals set up by individuals in decision-making positions within organizations. Household members have more or less formalized organizational projects together, and one such project might be "to live a good life". For achieving this goal on a regular basis in a modern society, at least some kind of income is necessary and thereby one project for household members is to work.

Individual and organizational projects intersect. For example, "work to earn a living" is an individual project, but it is also part of the household's organizational project "to live a good life". In addition, it depends on the existence of an organization providing jobs to pursue its organizational project(s). An organization's organizational project is in turn dependent on the capabilities of the individuals they employ. The goal for the company organization project might be "to produce and sell high-quality services". Figure 3.10 shows the intersection between

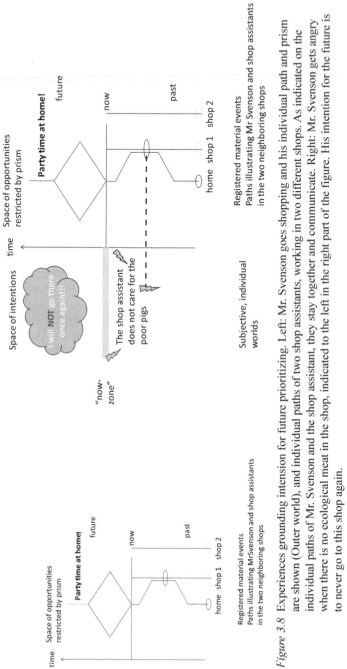

Figure 3.8 Experiences grounding intension for future prioritizing. Left: Mr. Svenson goes shopping and his individual path and prism are shown (Outer world), and individual paths of two shop assistants, working in two different shops. As indicated on the individual paths of Mr. Svenson and the shop assistant, they stay together and communicate. Right: Mr. Svenson gets angry when there is no ecological meat in the shop, indicated to the left in the shop, indicated to the left in the right part of the figure. His intention for the future is to never go to this shop again.

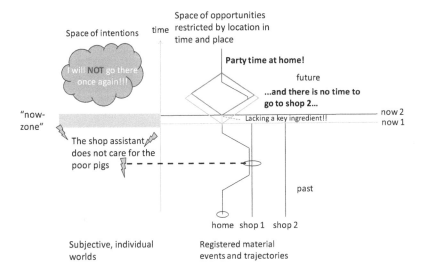

Figure 3.9 When the Inner world's intensions collide with the conditions of the Outer world. Mr. Svenson discovers that he has forgotten a key ingredient, and since his prism does now allow him to go to shop 2, he has to return to the shop he had intended to not visit again.

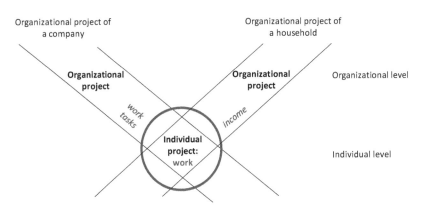

Figure 3.10 The intersection between two organizational projects and one individual project. At the organizational level there are companies and households, and they have different goals. The goals for the company are to make profit and produce high-quality goods. They need skilled employees who can perform the necessary work tasks. The overall goal for a household is to live a good life. A job and an income are important for adult household members. The individual project of an employee is to earn an income, and to do it she is employed by an organization, integrated in the production process and assigned to perform specific work tasks. The goal of the individual work project is to get resources for living a good life.

the household organization and the company organization and their respective organizational projects in the individual project, work.

Constraints

People use time to rest, care for themselves and others, and sleep in their home, and they leave home in the morning and return there towards the evening. This principle of return delimits the opportunities for the individual to move away from home, as indicated by the prism. The principle of return is a constraint, delimiting the individual's prism. People encounter constraints[14] in their efforts to achieve the goals of various projects, and there are a myriad constraints to handle as they navigate "now" in order to fulfill the goals.

To achieve the goals of many projects, people have to get in touch with other individuals at other places in the future. They also have to adjust to the time limits set by other individuals for certain activities. Hägerstrand (1970) concluded that it is not possible to classify all constraints in detail as time-space phenomena, but there are three large aggregations of constraints in time-geography: *capability*[15] (or *capacity*) constraints, *authority* (or *steering*) constraints and *coupling* constraints. The first two constraints resemble those usually discussed within social science and relate to living individuals.

Capability constraints concern the individual's opportunities related to her bodily and mental functions and to the resources that are available for her. Individuals with obvious capability constraints are children or very old, sick or disabled individuals. Limited command over tools and transportation are other examples of capability constraints. In human populations, capability constraints might also refer to, for example, slow means of transportation and a lack of economic resources or knowledge about how to solve a problem.

Authority or steering constraints concern power relations. In human populations, for example, children are subordinated to the authority of their parents and teachers, which makes children constrained. All citizens are subject to the national laws and have to obey them; otherwise they risk punishment. Other examples of authority constraints are created in hierarchical organizations where norms and rules are set up which exert authority over those subordinate to the people in power. In an organization, such as a school, a workplace or a service provider, there are rules that the participants or members are expected to obey; hence, the rules steer the activities of the people involved. Authority constraints also delimit who is allowed to get into a place or building. Authority constraints are products of culture, a Popper's World 3 construct. Steering constraints hinder individuals from acts that they might have strived for, and power may be exerted both by laws, rules and norms (Popper's World 3) and material objects (Popper's World 1).

The coupling constraints are deeply anchored in time-geography: they stem from people's opportunities and the need to couple and de-couple. In terms of elementary events, this concerns their need to move, arrive, get in touch, stay in touch, leave and disengage. Hägerstrand wrote about the individual's opportunities to perform activities located within reach of that individual: "inside the daily prism is to a pronounced degree ruled by 'coupling constraints'. These define where, when and for how long, the individual has to join individuals, tools, and materials in order to produce, consume, and transact" (Hägerstrand 1970: 14). For example, children need an adult staying in touch with them all day in order to be protected and to get care, help and guidance. Then the child's capability constraints result in a coupling constraint for the child as well as for the adult. Coupling constraints, then, relate to the two other kinds of constraints and concern the opportunities to establish time-space couplings between individuals (them getting in touch, in Popper's World 1).

Many employees find their work hours' scheduling to be an authority constraint. This constraint is easier to handle for a person who can decide when to start moving to the workplace, and how fast, than for a person who has to adjust to a public transit timetable. For the latter the bus introduces a coupling constraint when she tries to meet the authority constraints of the work time schedule. The person without a car must catch a bus at a certain time to arrive at work when expected. Thereby, the capability constraints are linked to and interact with coupling constraints. Then, both the bus timetable and the work schedule serve as authority constraints and result in coupling constraints.

Constraints are discussed by Hägerstrand (1970), Lenntorp (1976) and Mårtensson (1979), and many other later researchers have used the concepts.

Pocket of local order[16]

In his article about time-geography from 1970, Hägerstrand presented the domain concept and discussed its relation to authority constraints. He found the concept to be essentially spatial and suggested a redefinition of it "to refer to a time-space entity within which things and events are under the control of a given individual or a given group" (Hägerstrand 1970: 16). In the same article he also suggested that there is a hierarchy of domains, indicating that those who steer the superior domain also use their power to restrict what actions are permitted in subordinate domains.

The domain concept subsequently faded away in the time-geographic literature, probably because of its "essentially spatial" connotation, and was in turn substituted by Hägerstrand (1985) with the time-geographic concept pocket of local order. Pockets of local order are necessary for living individuals to carry out their projects. The pockets of local order are, if properly maintained, "sufficiently free from encroachment" (Hägerstrand 1985: 207) from processes and other projects in the nearby time-space. Living individuals might consciously arrange pockets of local order. Pockets of local order may also be arranged by nature from the local conditions in climate, soil, animal populations and vegetation.

Hägerstrand writes: "The human pockets of local order are a superstructure, directly added to nature and not possible to maintain without that base" (1985: 208). Such pockets of local order are exemplified by "arrangements in the landscape, homes, and factories" (Hägerstrand 1985: 208).

There are hierarchies of pockets of local order. They appear as Russian dolls wherein a smaller pocket of local order resides within a larger one. A home is a pocket of local order, inhabited and upheld by the household member(s) and their performing of individual and organizational projects. Each room in the home is also a pocket of local order, adjusted for specific projects, like the kitchen, drawing room or bedroom. As the household members change, a child is born or a child moves out of the family home, the pocket of local order changes. There is a need for a room for the child to play, or the room where the now-absent child resided is converted into a room for hobbies.

Hägerstrand's concepts and notation system recognize the intimate relationship between time and space. He said:

> So far continuity over time has been given prominence. This is not enough. The full grasp of processes requires that also space is viewed as a continuum. No holes are permitted through which entities and events can fall out and be forgotten. Time and space are intimately related. It takes time to cover distances, even the shorter ones. In addition space, understood as room, is a regulator of events because of its limited capacity to accommodate objects and thereby also processes. Competition over the same piece of space is one of the most common sources of conflicts. To view time and space simultaneously as continua, populated by elements having different probabilities of survival, ought to help us to detect the collateral characteristics of processes of change. We would be able to see more clearly how various strands move forwards in time through space shoulder by shoulder, so to speak, while they sometimes support and sometimes block each other.
>
> (1987: 49)

The following chapters present examples of how this mode of thinking, with its conceptual and visual tools, is applied and further developed in a wide range of thematic fields.

Notes

1 This is obvious in most of his texts after 1970.
2 For animals other than humans, the *program* concept is used to investigate their activities.
3 Also, animals may create pockets of local order, like their nests and territory, which they defend from intruders. However, as we understand it they do not consciously set up long-term goals like humans.
4 The time-geographic use of the *individual* concept was discussed by Hägerstrand in his first international article on time-geography in 1970, and he furthered the discussion in an article in Swedish in 1974.
5 There are some similarities with the actor–network theory. Here, the aim is to present the time-geographic approach and therefore the text is not mixed with other approaches.
6 The *path* concept was introduced in Hägerstrand (1970) and it is closely related to and constitutes the basis for developing the prism concept as elaborated by Lenntorp (1976). A synonym of *path* is *trajectory*.
7 There is an obvious thing, which is empirically revealed by using individual paths to illustrate the indivisible human individuals' daily life: most of the time people stay at the same place. Not surprisingly, the home is used for the longest time of the day and the activity with the longest duration, sleep, is performed there.

8 The time span of existence varies between different individuals and within different populations.
9 These elementary events for an individual path regarded as isolated from other individual paths are of course valid for each individual path when analyzed in combination.
10 Examples will be presented in the application part of the book.
11 The *prism* concept was applied in Hägerstrand (1970). It is thoroughly presented by Lenntorp (1976) and has been further developed by Miller (1991) and Miller (2005).
12 Hägerstrand discussed the *population* concept in Hägerstrand (1985).
13 This figure is presented in various forms in Lenntorp (1976), Mårtensson (1979) and Hägerstrand (2009).
14 The *constraint* concept was presented in Hägerstrand (1970).
15 Sometimes the concept of *capability* is used instead of *capacity*.
16 The *pocket of local order* concept was first mentioned in Hägerstrand (1985).

References

Hägerstrand, T. 1970. What about people in regional science? *Regional Science Association Papers*, Vol XXIV, pp. 7–21.
Hägerstrand, T. 1974. Tidsgeografisk beskrivning. Syfte och postulat. *Svensk Geografisk Årsbok*, 1974, pp. 87–94.
Hägerstrand, T. 1976. Geography and the study of interaction between nature and society. *Geoforum*, 7, pp. 329–344.
Hägerstrand, T. 1985. Time-geography: focus on the corporeality of man, society, and environment. In *The Science and Praxis of Complexity*, pp. 193–216. Tokyo, Japan: United Nations University.
Hägerstrand, T. 1987. On levels of understanding the impact of innovations. In *Gesellschaft – Wirtschaft – Raum. Beiträge zur modernen Wirtschafts- und Sozialgeographie. Festschrift fur Karl Stiglbauer*. M. Fischer and M. Sauberer (eds). Vienna, Austria: Arbeitskreis fur Neue Methoden in der Regionalforschnung.
Hägerstrand, T. 2009. *Tillvaroväven*. K. Ellegård and U. Svedin (eds). Stockholm, Sweden: Formas.
Lenntorp, B. 1976. *Paths in Space-Time Environments. A Time-Geographic Study of Movement Possibilities of Individuals*. Meddelanden från Lunds universitets Geografiska institutioner. Diss. LXXVII.
Mårtensson, S. 1979. *On the formation of biographies in space-time environments*. Meddelanden från Lunds universitets Geografiska institution, Diss. LXXXIV, Lund.
Miller, H.J. 1991. Modelling accessibility using space-time prism concepts within geographical information systems. *International Journal of Geographical Information Systems*, Vol. 5, No. 3, pp. 287–301.
Miller, H.J. 2005. Place-based versus people-based accessibility. In *Access to Destinations*. David M. Levinson and Kevin J. Krizek (eds), pp. 63–89. Amsterdam and Boston: Elsevier.
Popper, K. 1977. The Worlds 1, 2 and 3. In *The Self and its Brain*. Popper, K. and Eccles, J. (eds). New York, NY: Springer International, pp. 36–50.
Popper, K., and Eccles, J. (eds). 1977. *The Self and its Brain*. New York, NY: Springer International.

Part II
Applications of the time-geographic approach

4 Urban and regional planning

The societal context of the development of time-geography in Sweden

The transformation of Swedish society during the 20th century influenced the life of Torsten Hägerstrand, as indicated in Chapter 1. During his childhood, most people still lived in the vertically linked, short-distance society, but by the time he was a teenager the situation had changed and the horizontally linked, long-distance society gradually overtook the vertically linked society (Hägerstrand 1970b, 1988). The following section gives the contours of the development of the Swedish society, which serves as a background to Hägerstrand's contributions to urban and regional planning, and relates closely to the developments of the time-geographical approach.

In the early 20th century, Sweden was a poor country, with low living standards. However, there were strong desires to change the situation, and efforts were made both by industrialists and political movements to improve the situation for the population. In 1928 the leader of the Social Democratic Party, Per Albin Hansson, launched the idea of the Swedish *Folkhem* (home for the people), and the subsequent realization of it established this party's long-lasting power in government. The reforms of the *Folkhem* period were the basis of the modern Swedish welfare state and it was not until 1976 that the social democrats lost their governmental power.

Urbanization and industrialization brought striking changes to Swedish society and through the 20th century a steady stream of migrants left the rural life, hoping for better work and a life in urban areas. Viewed from a longer historic perspective (see Figure 4.1), the proportion of the population living in urban areas changed from only 20% in 1880 to an equal share with rural areas in 1930, and by 1970 about 80% of the population lived in urban areas. The turning point from a rural to an urbanized society was passed in the late 1920s.

The journalist and author Lubbe Nordström traveled all through Sweden during the 1930s and reported about the conditions of the people he interviewed (Nordström 1938). He reported that living conditions, housing, wages and work in the countryside were not good for the majority of people. He commented on families living in houses with earthen floors, bad insulation and, of course, no

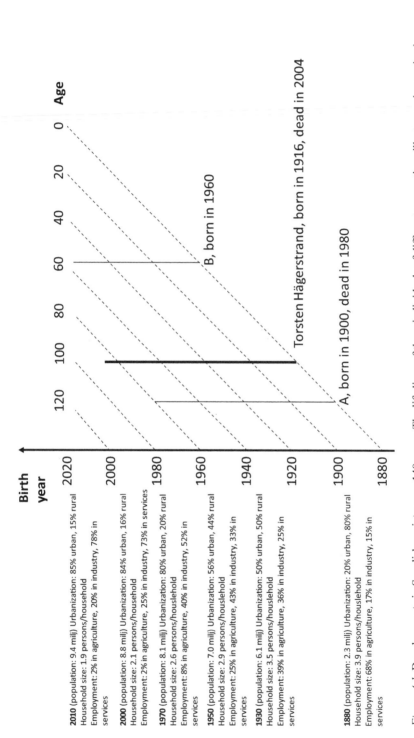

Figure 4.1 Developments in Swedish society over 140 years. The life lines of three individuals of different generations illustrate what societal changes they experienced during their life. For the oldest one, A, urbanization was a great adventure, while B was born into an urbanized country. The life line of Torsten Hägerstrand is also illustrated.

central heating, sewage facilities or running water. He was concerned about the unhealthy living circumstances, especially for the children. Nordström was one of the advocates for industrialization and urbanization and he described the high chimneys with smoke from modern factories as a sign of a prosperous future.

Such were the living conditions when social democratic leader Per Albin Hansson presented the idea of the Swedish *Folkhem*. Based on democratic values, schemes aiming at a good home for everybody were set up by politicians and plans for their achievement were prepared. Several reforms were eventually put into practice, making life easier for the majority (see Figure 4.2). For example, basic pension reform was realized in 1935; child benefits were introduced in 1937 – they were initially based on means tests, but from 1948 there were benefits for all children; a two-week general holiday reform for all was put into practice in 1938; and a public system for general health insurance started in 1955. The reforms made considerable differences and helped found the Swedish welfare system that eventually matured during the 20th century. The *Folkhem* policy included improved material living conditions, especially as regards housing and electrification, and the intention was that every citizen should live in a high-quality home equipped with modern conveniences like a bathroom, central heating, running water, electric stove, and so on. Based on such ideals, many new city districts in the rapidly growing urban areas were constructed, and some of these are still attractive today because of the high-quality architectural solutions and good materials and construction.

The development of the *Folkhem* was put on hold during the Second World War. Since Sweden was not directly involved in the war, the national economy quickly recovered after the peace and the creation of the modern welfare state could continue. A dominating term for the Swedish situation during the period from the end of the war until the first global oil crisis in 1973 is "the record years", characterized by increasing industrial production and steady economic growth. The urbanization continued and there was a big transformation of the Swedish labor market. From the 1930s until the mid-1960s the largest portion of employees was found in the production industry sector, but from about 1970 the service sector became dominant (see Figure 4.1) and the migration to urban areas continued.

To meet the demand for housing in the urban areas a national program to build one million homes in a 10-year period was launched in the mid-1960s (see Figure 4.2). This program was also a way to improve the housing standard. The policy to improve dwelling conditions for the majority yielded results. During the period from 1880 to 2010 the average household size decreased substantially, from 3.9 to 1.9 members per household, and there was little overcrowding in Swedish homes (see Figure 4.1).

A large portion of the million-program housing was constructed in new suburban areas, but centrally located areas were also refurbished. The latter caused controversies. Many old, small and unmodern apartment buildings in the central parts of the cities were torn down to create room for new constructions in these attractive locations. In the 1960s and 1970s young people were strongly opposed to the complete destruction of old housing in the central parts of towns. They had

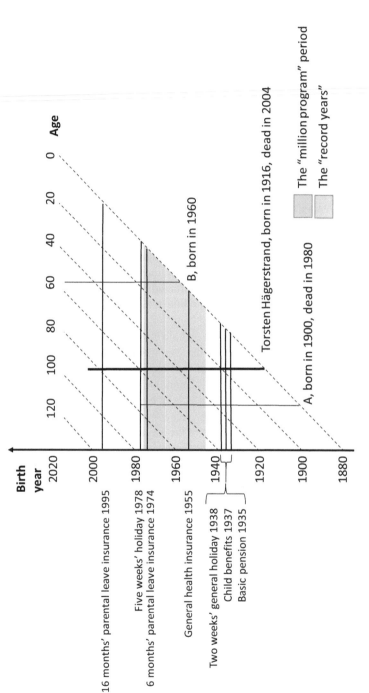

Figure 4.2 Reforms that framed the development of the Swedish welfare society. The "million program" reform for building one million homes lasted for 10 years (blue), and the "record years" were characterized by exceptional economic growth for about 15 years (orange).

not experienced the shame of the family living in a one-room apartment in a poor neighborhood with an outhouse and just cold water in the kitchen. When this post-war generation left their parents' homes to live on their own, many wanted to live close to the city center and move into apartments with low rents. The centrally located old, low-standard apartments were attractive to the young but detested by the older generation of social democrats who had grown up there. The younger generation also opposed the standardized apartments in the massive and industri-ally constructed million-program-home areas in new suburbs.[1] They claimed that even if the material living standard increased, a good life did not automatically follow. Another critique of the million program concerned the functional planning of the new suburbs, which separated homes from workplaces and service areas, resulting in long distances to workplaces and services.

Around 1930 emigration from Sweden equaled immigration, and the total number of people moving across the border was low. During and after the Second World War immigration increased and it has since been substantially higher than emigration. Beginning in the 1950s, many labor immigrants were recruited to work in the rapidly growing Swedish industrial production sector. There was still a lack of labor and from the 1970s female participation in the labor force was encouraged. Some important actions were decisive for women's entrance to the labor force. One was the introduction of high-quality public childcare, and another was the Swedish system for paid parental leave, which was introduced in 1974 (see Figure 4.2). In combination with a growing labor market welcoming female employees, this reform freed many married women with children from economic dependence upon a male family wage earner.

The rapid economic growth was based on the success of Swedish industries' production and sales, which built on a secure supply of cheap energy, both oil products and domestic electricity. From the mid-20th century, the government set up plans for national investments in infrastructures for the growing automobile traffic, secure energy supply, housing and the location and organization of health care and public services. Many of the national political reforms initiated from the 1930s until the 1970s aiming to increase welfare for the population were under-pinned with national studies.

From the 1950s, several of these national investigations involved social science researchers of the same generation as Torsten Hägerstrand. Human geog-raphers, economists and sociologists were engaged for studies regarding many policies. Hägerstrand was engaged in investigations concerning, for example, urban and regional planning, regional policies and policies for natural resources (Hägerstrand 1966; Hägerstrand et al. 1967; Hägerstrand 1969; Hägerstrand and Öberg 1970; Hägerstrand 1970c; Hägerstrand and Lenntorp 1974); consequences of constructing a bridge from Sweden to Denmark (Hägerstrand 1967); methods for a new municipal structure (Hägerstrand and Godlund 1961); national parks (Hägerstrand 1991); natural resources (Hägerstrand 1988); and future studies (Hägerstrand 1972 and Myrdal et al. 1972). He also served as an expert in inves-tigations concerning road transport and planning of road networks at regional and national levels.

The urbanization and industrialization affected the development of the *Folkhem* in many ways. Urbanization meant that people left rural areas in favor of larger population centers, and the costs for supplying people with welfare services in the depopulated districts increased. A geographic centralization of public service supply was inevitable if taxes were not to become too high. In the 1970s efforts were made to move some national public services from the capital to other cities in order to hamper the concentration of jobs. Investigations to see what could come out of such relocations involved social science researchers, and some of them used time-geographical arguments to suggest locations of public services (for example, Öberg 1976; Mårtensson 1979).

The changed share of employees in the three economic sectors (agriculture, the production industry and services) reveals the development towards 'de-agriculturation' (see Figure 4.1), transforming Sweden into a horizontally linked, long-distance society. In 1880, 68% of the population was working in the agricultural sector, while in 1970 only 8% of the employed population worked therein and in 2010 the corresponding figure was as low as 2%. This development, of course, had effects on people's relationship with nature and the landscape and thereby also on their knowledge of, for example, how food is produced and handled before it is brought into the urban households' kitchen table (see Chapter 1). Industrialization and urbanization processes led to new generations of Swedes growing up without immediate experience of or even basic knowledge about the relationships between human life, cultivation of farmland and the production of raw materials for food. Industrialization included a growing food industry, which was a prerequisite for feeding the increasing urban population. However, the link between humans' livelihoods and nature based on everyday experiences was broken for an increasing proportion of the population, and Hägerstrand was worried about this development into what he called the horizontally linked society (Hägerstrand 1970b).

Hägerstrand in the society in transition

Sweden shifted from being an emigrant country in the 19th century into a country with increasing immigration in the 20th century.[2] During his PhD studies Hägerstrand was sent out by his professor to study what happened in a region from which many emigrants had gone to America in the 19th century (see Chapter 2). From this experience, he gained an understanding of how migration was fueled by poverty and people's hopes for better work and living conditions at another farm, in America or later on in production and the cities. Hägerstrand's first publications are in the migration field; there is an article on the migration in Asby parish from 1947 (Hägerstrand 1947) and another on the livelihood of crofters in the Asby parish in 1950 (Hägerstrand 1950).

Not only emigration from Sweden but also immigration to Sweden played an important, but very different, role in the academic career of Torsten Hägerstrand and his development of the time-geographic approach. One refugee coming to

Sweden after the war was the Estonian human geographer Professor Edgar Kant. He was employed by Lund University; however, initially not as a professor.[3] Despite his formal position on the academic fringe in the beginning of his life in Lund, he became an important inspirational source and discussion partner for Torsten Hägerstrand. Kant brought with him new directions in human geography, inspired by, among other Central European geographers, Walter Christaller's central place theory and theoretical approaches building on quantitative methods. Both Kant and Hägerstrand found the dominating regional geography too descriptive and atheoretical. The discussions with Kant encouraged Hägerstrand to work with both large empirical data sets and quantitative methods in his dissertation about innovation waves (Hägerstrand 1953/1967).

The initial research work on migration as a general phenomenon, where Hägerstrand followed the indivisible individuals as they moved from home to home, was empirically based on material from Asby parish, such as maps, church registers and other public registers (see Chapter 2 and figures 2.6 and 2.7). It was a wearying task to find and compile all the data. Hägerstrand concluded that it would be of great help to have a national register of all real estate properties, including their geographical coordinates. Such a register would also make comparisons over time easier, since such research investigations would no longer depend on administrative borders. The problem is that administrative borders change depending on the needs of the public sector administration, with no respect for the needs of researchers. Hägerstrand, therefore, suggested such a national register for real estate properties would be useful for both administrative and research purposes, based on a technology called "data processing machines" (Hägerstrand 1955). This article is a forerunner of what later became the development of geographical information systems (GIS) and the Swedish national register of properties based on the coordinates of their location.

In the early 1960s Hägerstrand wrote an essay on life and production on a large mansion in Skåne, Svaneholm (Hägerstrand 1961a), where he reflected on the landscape formations and nature, and how the landscape changed depending on the organization and performance of agriculture. Herein, he reveals deep knowledge about the conditions for agriculture, and a bit of nostalgia. The experience in the flesh of the big shift from a society dominated by agriculture, forestry and small industries into a service-dominated, urbanized society probably influenced Hägerstrand in developing the ecologically based orientation of the time-geographical approach, which was first articulated in the late 1970s and increasingly so from the 1980s. In 1988, Hägerstrand made a thorough investigation of the national regulations and laws that lay behind the transformation of the Swedish landscape (Hägerstrand 1988). Thereby, he made the authority constraints set up by society on the development of the landscape obvious. He was worried about the increasing segmentation of food production and its time-space detachment from eating, and the lack of knowledge about the consequences of the conditions under which food is produced, not least the wider effects on the ecosystem.

Hägerstrand's involvement during the 1960s and 1970s in national investigations for the government and parliament concerning regional policy and urbanization resulted in the initiation of a governmental research institute on regional development, ERU (*Expertgruppen för regional utveckling*). This institute did investigations and published research reports in order to improve Swedish regional policy. A related book, *Urbaniseringen* (Hägerstrand 1970b), included chapters written by several of the most influential Swedish social science researchers of the time. Among other things, Hägerstrand wrote about the transition of Sweden from a vertically to a horizontally organized society (Hägerstrand 1970b).

One national investigation on regional policy that involved the entire time-geography research group in Lund was published in Hägerstrand and Lenntorp (1974: 2). Here, the group made a theoretical contribution based in time-geography with a focus on how to match local supply and demand for public services. In matching the local supply and demand, the indivisible individual was at the core and towards the background of the principle of return (prism); for example, the possibility for household members to perform different "daily activity programs" was investigated and the effects of their possession of various means of transportation (car, public transport, bicycle) were simulated.

An additional theoretical contribution emanating from Hägerstrand's partaking in national investigations is a report for the government's Future delegation led by minister Alva Myrdal. Here he showed the importance of the historic context preceding the current situation when doing future studies. Time-geographically, this perspective on future studies was underpinned with visualizations of indivisible individuals in the context of societal changes during their lifetime (Hägerstrand 1972). This principle is the basis of figures 4.1, 4.2 and 3.5). It gives a contextual understanding of events which helps interpret the effects of what people experience during their life course when previous futures are transformed into the past. The point of departure is the lifetime of the indivisible individual, regarded in the context of societal change over time. Hereby, the generational effects of, for example, reforms or major societal events are made clear. Related to this is the publication of the book *The Biography of a People*, in which the population of Sweden is related to the historic and future development of society (Guteland et al. 1974).

As the 20th century went on it became more and more obvious that regional policy could not overrule the economic power of enterprises and, consequently, the location of jobs. Hence, the de-population of rural areas continued, leaving the elderly and the few engaged in the agricultural sector behind, while the biggest cities grew. Services were lacking in the rural areas. Hägerstrand was disappointed by the limited effects of his efforts to contribute to the shaping of how to think when formulating policies and creating the means for their realization. The policy makers did not recognize, or maybe did not understand, the importance of looking at the individual as an indivisible over time (Hägerstrand and Lenntorp 1970).

Developing the time-geographic approach from studies on urbanization

A time-geographic take on supply and demand for services

Hägerstrand's research group in Lund grew and, as mentioned above, over time it was increasingly involved in the government's and parliament's national investigations. The group aimed at a deeper understanding of the ongoing urbanization process and identifying new approaches to urban and regional planning so that all citizens, irrespective of their home location, could gain from the welfare society and the public services. The time-geography researchers tested the time-geographical framework and thereby they brought the daily life of people into planning. In the approach, people are handled as indivisible individuals with projects and efforts to achieve a good daily life for themselves and their households. Consequently, people are not just numbers. This is the implication of the title of Hägerstrand's article from 1970, "What about people in regional science?" (Hägerstrand 1970a). The time-geographic take on urban and social planning commenced with studies of the differentiated regional location of people in Sweden and, following on from these differences, the variation in opportunities for the individuals to reach daily services.

The interdisciplinary research project "The urbanization process", which gave Hägerstrand financial opportunities to build his research group in Lund (see Chapter 1), can be seen in the light of the need for more knowledge about the societal changes emerging from urbanization. The group presented its time-geographic research, including its core concepts, in an official government investigation (Hägerstrand 1970c). Hägerstrand elaborated on the prism and thereby widened the planning perspective from merely the geographical distance to a time-space distance for analyzing people's available opportunity space for performing activities of importance for them. Hägerstrand and Öberg (1970) investigated empirically the question of *how far and for how long* people had to travel in order to reach various kinds of service. Their analysis showed that the supply of services to people in urban areas was good, but that there were large regions outside the main urbanized areas where these services were not available within decent time-frames. They also found that, due to poor public transport organization and traffic congestion, the time people had to spend on commuting over short distances in the bigger cities could be as long as in rural areas where commuting over much longer distances took place (Hägerstrand and Öberg 1970). This was important knowledge in a society undergoing increasing centralization and urbanization with political ideals of equal opportunities for people in the whole country, irrespective of their home location.

In the late 1960s, the expanding production industries demanded labor and many Swedish women were encouraged to entered the labor market. There was a discussion about demand and supply of labor, based on an understanding of individuals as numbers without paying attention to neither humans' indivisibility and their social and geographical contexts, nor to the time dimension. Hägerstrand (1970c) presented an alternative way of thinking about the relationship between people and the labor market based on time-geography, with its basic assumption about the indivisible individual who has to perform some daily activities in a specific sequence in order to achieve the goals of their projects (see chapters 2 and 3).

In Figure 4.3, an individual path illustrates an indivisible person's movements in the time-space, between some geographic locations of stations she has to visit during this day (home, workplace, bank and post office) and their opening hours. In the early 1970s, this kind of illustration was new and it challenged the traditional way of thinking since it highlighted the necessity to take both time and place into overt consideration when planning for and deciding about the location of homes, workplaces, services and the time-scheduling of work, services and public transportation.

Hägerstrand (1972) presented an approach to better understand the matching problem between people and the labor market, the "activity system" model of society. This activity system is highly abstract and consists of two main parts: the "activities" to be performed (in public and private business organizations and households) and the individuals in the population with their time. The population time includes the time to be lived by all individuals in the whole population, with their different ages, education, occupation, location, and so on. Hence, this is a wider perspective compared to the traditional perspective of supply and demand in the labor market wherein the youngest (children and adolescents) and oldest people (retired) are often neglected since they are regarded as "unproductive". By recognizing a generational perspective (compare figures 2.5, 4.1 and 4.2), the time-geographic model considers the indivisible individual and, thereby, that old people have previously contributed to the development of the society and that the young

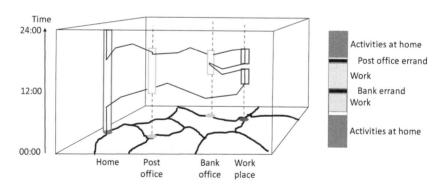

Figure 4.3 An individual's daily movements from home and back in a local settlement including visits to a post office, bank office and workplace. To the left: the individual path in the time-space context. To the right: the activity sequence performed by the individual. (Modified from Hägerstrand and Lenntorp 1974: 227.)

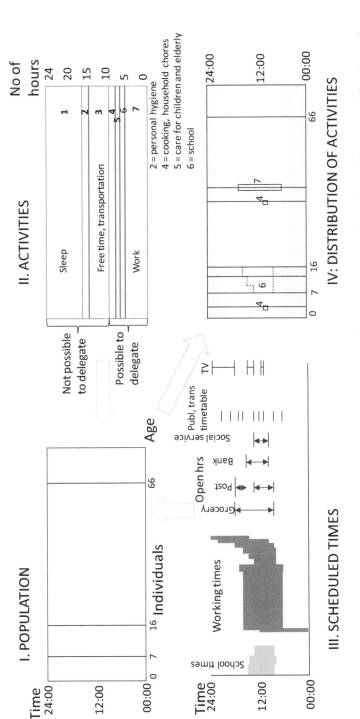

Figure 4.4 Matching individuals in the population with activities, under constraints of organizations' timetables and human needs for sleep and rest. The population time is illustrated in the upper-left part: the lines indicate a person that is 7 years old, a person 16 years old and a 66-year-old. In the upper-right part, the population time is filled with various activities to be performed by the individuals within the constraints of the available population time. Observe that the activities in this part are not distributed over the individuals. The lower-left part of the figure shows the timing of organizations' demand for individuals in terms of when they can (or must) spend time on activities in schools and workplaces. Opening hours in services and timetables for public transit are also illustrated. Finally, in the lower right activities are distributed over the indivisible individuals in the population, showing for what purpose each person uses her time. In this figure the time use for various activities during the day of the 7-, 16- and 66-year-old people from the upper-left part is shown. The dotted area indicates sleep in the early morning and late evening. (Modified after Hägerstrand 1970b: 147.)

people will do so later in life. Then, all people either will have been, are or will be in the "productive" ages, and with such an approach the young and the old should be seen as equally important as those in-between, currently in the workforce.

The micro-level example presented in Figure 4.3 was the basis for construct-ing a model of the activities occupying the whole population on a daily basis (see Figure 4.4). Herein, the available population time (each day the population time equals 24 hours x the number of individuals in the population) is divided into rough categories of activities that people must perform to live and reproduce the society. There are necessary activities (sleep, personal hygiene and meals) which all individuals have to spend time on, and time for transport – which cannot be delegated to someone else. Then, there are activities that might be performed by another individual (delegated) like cooking, care for the home, care for others, work and school. However, somebody has to do it. All population time is com-pletely used up every day. The population time, then, is regarded as a kind of "supply of time" that can be "demanded" from public and private organizations in their business to withhold and develop the society.

Access to services in central and remote parts of a municipality

Mårtensson (1974) analyzed individual- and household-level effects of the loca-tion of a home. She made refined analyses of individuals' opportunities to find and utilize services' supply according to the requirements of their "daily programs". She was interested in differences regarding people's opportunities to perform activities depending on where they live and what means of transportation they are in charge of. She found big differences in people's opportunities to reach vari-ous kinds of services depending on their location in a region and what means of transport they commanded. Mårtensson gave principal examples of the variation in the time needed for medical treatment, fixed between 3pm and 3.30pm, in the different settlements in Örebro municipality. On one hand, a person with access to a car could fulfill the medical care treatment errand within a time period of about one hour, irrespective of which part of the community she lived in. On the other hand, a person who lived in a peripheral part of the community and who had to go by public transportation had to use a substantial part of the day to get to the doctor at the same fixed time for treatment. This analysis includes the time for treatment, waiting and the transportation to and from the doctor's location, and this is of fundamental importance, since the treatment in itself takes the same amount of time for all, irrespective of where they live. An individual living far away from the service supply and with no car available for transportation had to start the trip at about 11.30am and could not return home until after 5pm. However, for such a person neither the time used for the transportation in itself, nor the time for treat-ment, was very long, but because of the bus timetable this individual had to spend a long time waiting, first until it was time to meet the doctor and then for the bus to return homewards. Mårtensson's exercises showed that the regime for making up the timetables of the public transportation was built on assumptions that were not anchored in, and therefore did not fit, people's daily life activity programs.

Today, most public transportation organizations in Sweden adjust the timetables to working hours, which was not the case at the time of the studies presented above. In addition, the legislation on when shops, restaurants and other services are allowed to be open has changed over time. This is a necessity since now most individuals in the adult population are active in the labor market, and it would not be possible to maintain the same limited opening hour regime as in the mid-20th century. Whether this is a result of insights from the research or an outcome of the experience, the obvious mismatch between a livable daily life and the traditional organization of the timing of work and service activities can be discussed.

Later, the problem for rural areas of supplying services prescribed by law to a stagnating or decreasing population was addressed in the dissertation of Cedering (2016). She used time-geography to investigate the consequences on daily life activities and social contacts for children in rural areas where the local school was closed. After the centralization the children had to travel further to get to school, their social networks changed and they became more dependent on their parents driving them by car to their new friends since they lived farther away. Cedering uses time-geography to investigate the wider social time-space consequences of the closure of schools, and bases her analysis on the results of studies of families' new daily movement patterns in the municipality. She concludes that, apart from the long-term consequences of the school closure, an immediate product is that the locally organized activities became less frequent since people, both adults and children, increasingly had to spend time on the move.

Power in pockets of local order in urban and regional development

Wihlborg (2000) investigated the consequences of the Swedish national policy for introducing information and communication technologies for people living in rural areas. She used the time-geographic concept *pocket of local order* at the municipal level, showing that the conditions for people and companies in small rural municipalities to utilize the new technologies relied very much on local initiatives. They had to set up an order themselves to introduce the material parts of the technologies and to organize courses to learn how to use them. Her results show that even though there is a policy that all parts of Sweden should have internet connections, there are still big differences. There is little interest from the big companies in investing in infrastructure in rural areas, and of course the local collaboration between citizens to solve the problems takes a lot of time and effort, but it also helps in keeping up the local sense of belonging to a community. This study concerns the first wave of introducing information and communication technology to the Swedish society.

In the early 21st century, the Swedish government set the goal that Sweden should be at the forefront as an information society. Thereby, infrastructures for the new generations of communication technology networks had to be constructed and spread over the whole country, as the political goal was that people in any location should have access to the new services. In a study about establishing the new-generation telecom networks, Wihlborg and Palm (2008) investigated the two different processes of constructing broadband and 3G telephone networks respectively. In the study, Wihlborg and Palm regarded the municipalities and the nation state as pockets of local order at different geographical levels. In pockets of local order, power is exerted. The national pocket of local order exerts power over the local municipalities, but it might be problematic to use the same kind of power to steer all municipalities since they differ; for instance, depending on whether it is a rural or an urban municipality. However, the cohesive nature of social networks empowered people in the rural municipalities as pockets of local order, giving them power from below. The national level is usually not familiar with the diversities at the local level, especially not in remote rural areas. Due to the local anchoring of the broadband in the municipalities, the process went rather smoothly and local organizations were engaged in building the new infrastructure. Especially in remote municipalities, other social networks were also enrolled, and the order of the introduction of broadband in the municipality as a pocket of local order was agreed upon. Contrary to the locally anchored process introducing the broadband infrastructure in rural municipalities, the introduction of the 3G net driven by big commercial companies met resistance from people in the local municipalities during the construction phase. This was especially evident as regards the geographic location of the tele-masts, which constitute the material part of the 3G net. Further, the big companies did not find it profitable to establish the net in remote areas. The analysis based on the pocket of local order concept anchors complex processes of policy making in the local material geography (Wihlborg and Palm 2008: 49). One core conclusion of the study is that policies concerning construction of infrastructures and other socio-technical systems should be grounded in the local physical and social contexts. Then the systems become policy issues of the pocket of local order.

Time-space location of home, work and childcare service in Japan

The Japanese society is modern and industrialized and has world-leading industrial conglomerates, but it is still traditional as regards the ideal of males as breadwinners. Today, many Japanese women are well educated and want to make a career over their lifespan. However, the traditional culture of male breadwinning and female housekeeping and responsibility for childcare forms authority constraints, hindering women's plans to work in career professions if the home is located far from the workplace. Time-geographic studies carried out in Japan show that the time-space organization of limited nursery service supply has severe repercussions for family life and career opportunities for women (Okamoto and Arai 2019).

A Japanese research group with geographers inspired by time-geography investigated problems of mismatch between work schedules, public transportation timetables and day care for children in Japan (Kamiya et al. 1990). They found from their time-geographic analysis of mothers' opportunities to participate in the labor market that there is a need to improve the supply of nursery services. Different forms of childcare were included in the study, apart from nursery school, such as care given by grandparents. The fixed times at nursery schools together with the short opening hours, forcing parents to pick up children at a certain time, hinder mothers from working full time. The researchers called these fixed times for leaving and picking up "markers" of the day, between which there is an opportunity space for work (or other activities) (Kamiya et al. 1990). Another time-geographic concept involving mothers' leaving and picking up of children is coupling constraints, closely related to the authority constraints regulating the time-schedules. There are several possible reasons behind the mismatching problems; for example, women usually do not use a car for commuting, public transportation takes a long time and the timetables do not correspond well with working times, and it is difficult to find a suitable workplace close to home. Altogether, this hinders women in making a career. The researchers argue for day-long nursery care as an effective way to help women to reach their goals with their work project (Kamiya 1999).

Chai (1993) clustered Japanese men and women according to their time-use pattern, and showed that women spend considerable time doing housekeeping. In that study, different patterns of utilizing the city space also appeared, depending on whether a person's home location is in a suburb or in the city center. On weekdays, people living in the city center stay in the center, while the activity space of people living in suburban areas includes the city center as well as the suburb.

In the wider context of urbanization and changes in the labor market, Okamoto studied the daily life conditions of couples in suburbs (Okamoto 1997). He showed how the time-space organization of work (for men and women), childcare, location of homes and workplaces, and transport mode influence the living conditions and indirectly also birth rates in Japan. Kamiya (1999) explained the problems of low female participation in the labor force and low birth rates in Japan by analyzing data from time diaries and day care service provision. He recommended policies such as extended day care opening hours and relocation of day care service to railway stations, and of offices to suburbs. Okamoto (1997) and Nishimura and Okamoto (2001) relate the declining birth rate in Japan to the culturally rooted division of labor in families, where women are responsible for almost all domestic chores. Okamoto and Arai (2019) suggest a way of solving the problem of locating affordable and accessible childcare, and base it on unconventional, time-geographically-grounded principles for transportation related to work and childcare. Their suggestion is to locate "nursery stations" close to railway stations, where people can leave and pick up their children, and the children are transported to a nursery school located further away from the railway station and spend their day there. When it is time for the parents to pick up the children, they are transported back to the nursery station, and the parents can pick them up on their way home.

Urbanization and restructuring the home–work relationship in modern China

To some extent, the rapid urbanization in China resembles that which the Swedish time-geographic researchers experienced in the early 1970s, even though the Swedish case is from an earlier date and on a dramatically smaller scale. The Chinese transition is investigated by Chai (2013) and his research group. Chai discusses the question of how urban and regional planning can take the complexity of everyday life of people into consideration when urban areas are in rapid transition in a country where processes of central planning and marketization are combined. He discusses methodological and theoretical challenges to research on urban developments in China and finds space-time behavior research, based in time-geography, important for understanding the complex and diverse transition in Chinese cities. Chai encourages further theoretical developments in the field.

In the transition of China into a market-like centrally planned economy, housing areas based on market conditions are erected and suburban areas are established on greenfield land and they grow quickly. The transition also influenced the old urban structures in which "Danwei compounds" were central elements. Danwei compounds were often located in central parts of urban areas and within them homes, services and workplaces were located in close proximity. One problem for people in most Western countries is to overcome long-distance commuting, measured in both time and geographical distance, between work, services and home. Contrary to this, the Danwei organizers aimed to locate homes, services, workplaces and leisure activities within the same local areas. Danwei compounds were "gated", with only a few entrances, and the life of the inhabitants was under control of the local authorities and the freedom to choose a workplace within the community was constrained. However, the Danwei compounds do not fit the current rapid urbanization and marketization, not least since the factories in the centrally located Danwei compounds create problems with air pollution and related health problems in the population and, in addition, the aggregate pollution contributes to climate change. Therefore, decisions were taken to transform and disintegrate the compounds. Thereby, a process of moving factories to less centrally located industrial areas was initiated, and the Danwei compounds lost their original multiple functions. In today's China, Danwei compounds are open, have more entrances and exist in parallel with market-based modern housing neighborhoods. Thus, on one hand inhabitants lost their work close to home, but on the other hand, they need less time for going from the community and out into the surrounding areas (Wang et al. 2011).

Wang and Chai (2009) show in a comparative study of transport demand for people living in Danwei compounds and in modern market-oriented suburbs respectively that in terms of commuting, the inhabitants of market-oriented neighborhoods spend much more time in transit, and they are increasingly expected to choose motorized means of transportation in the future. This development might counteract the intention to decrease air pollution, which was one reason to move factories from old Danweis. Their conclusion is that "Policy makers may need to look for policy measures such as land-use regulations and planning tools to retain the merits of the Danwei system in matching jobs and housing opportunities and containing transport demand" (Wang and Chai 2009). In the article titled "From socialist Danwei to new Danwei: a daily-life-based framework for sustainable development in urban China", Chai (2014) argues for a daily-life-based framework, using the time-geographic space-time activity concept. One challenge for time-geographers would be to analyze the material construction and social order of both market-oriented neighborhoods and modern Danwei compounds as pockets of local order.

In retrospect, and from the examples given in this chapter, Hägerstrand's time-geographic approach inspired many researchers to consider problems of relevance for local environments.

Hägerstrand argued that research should make a difference and support a sustainable life environment in society as a whole. Even though most people live in urban areas, they depend upon people living in smaller municipalities and rural districts. He said that social science disciplines have taken on problems at levels "far from the local and regional dimensions where most of the physical planning problems finally make a difference. A new orientation is necessary, not least for the sake of social sciences themselves" (Hägerstrand 1961b: 66, my translation).

Notes

1 There was a generational gap, and its emergence might be explained by thinking of it in terms of Figure 3.5, displaying the different experiences of people of different generations. It also relates to the concrete reforms as illustrated in figures 4.1 and 4.2.
2 Immigration was for the first time larger than emigration in 1930 and there has since been a steady immigration gain.
3 First Kant worked at the archive of the Geography department. Later he was employed as a lecturer and in 1964 he became Professor in Geography.

References

Cedering, M. 2016. Konsekvenser av skolnedläggningar. En studie av barns och barnfamiljers vardagsliv i samband med skolnedläggningar i Ydre kommun. *Geographica*, 8. Uppsala University, Department of Social and Economic Geography. Diss.

Chai, Y. 1993. Daily activity space of Hiroshima citizens: a case study of the citizens in their forties. *Japanese Journal of Human Geography*, 46, pp. 351–373.

Chai, Y. 2013. Space-time behavior research in China: recent developments and future prospects. *Annals of the Association of American Geographers*, Vol. 103, No. 5, pp. 1093–1099.

Chai, Y. 2014. From socialist Danwei to new Danwei: a daily-life-based framework for sustainable development in urban China. *Asian Geographer*, Vol. 31, No. 2, pp. 193–190.

Guteland, G., Hägerstrand, T., Holmberg, I., Karlqvist, A., and Rundblad, B. 1974. *The biography of a people: past and future population changes in Sweden, conditions and consequences*. Stockholm, Sweden: Allmänna Förlaget.

Hägerstrand, T. 1947. En landsbygdsbefolknings flyttningsrörelser. Studier över migrationen på grundval av Asby sockens flyttningslängder 1840-1944. *Svensk Geografisk Årsbok*, Vol. 23, pp. 114–142.

Hägerstrand, T. 1950. Torp och backstugor i 1800-talets Asby. In *Från Sommabygd till Vätterstrand*. E. Hedkvist et al. (eds). Linköping, Sweden: Tranås hembygdsgille, pp. 30–38.

Hägerstrand, T. 1953/1967. *Innovationsförloppet ur korologisk synpunkt*. Lund, Sweden: Gleerupska Universitets-bokhandeln. Translated into English by Allan Pred as *Innovation Diffusion as a Spatial Process*. Lund, Sweden: C.W.K. Gleerup.

Hägerstrand, T. 1955. Statistiska primäruppgifter, flygkartering och "data processing"-maskiner. Ett kombineringsprojekt. *Svensk Geografisk Årsbok* 1955: 233–255.

Hägerstrand, T. 1961a. Utsikt från Svaneholm. *Svenska Turistföreningens årsskrift*, pp. 33-64.

Hägerstrand, T. 1961b. Vem skall planera? In *Om Tidens vidd och tingens ordning*. G. Carlestam and B. Sollbe (eds). Texter av Torsten Hägerstrand. Byggforskningsrådet. T:21, 1991.

Hägerstrand, T. 1966. Regionala utvecklingstendenser och problem. Urbaniseringen. Statens Offentliga Utredningar (SOU) 1966:1, pp. 273–290.

Hägerstrand, T. 1967. Beräkningar rörande de olika Öresundsförbindelsernas inverkan på orternas "lägesvärden" i östra Danmark och södra Sverige. Statens Offentliga Utredningar (SOU) 1967:54, pp. 251–270.

Hägerstrand, T. 1969. Utvecklingsdrag som nödvändiggör regionalpolitisk insats. Appendix till Statens Offentliga Utredningar. Statens Offentliga Utredningar (SOU) 1969:27, Länsplanering, pp. 152–269.

Hägerstrand, T. 1970a. What about people in regional science? *Regional Science Association Papers*, Vol. XXIV, pp. 7–21.

Hägerstrand, T. 1970b. *Urbaniseringen – stadsutveckling och regionala olikheter*. Gleerups. Lund. Bröderna Ekstrands Tryckeri AB, Lund.

Hägerstrand, T. 1970c. Tidsanvändning och omgivningsstruktur. Statens Offentliga Utredningar. Statens Offentliga Utredningar (SOU) 1970:14, bilaga 4, pp. 4:1–146.

Hägerstrand, T. 1972. Om en konsistent individorienterad samhällsbeskrivning för framtidsstudiebruk. *Ds Ju* 1972:25. Specialarbete till Statens Offentliga Utredningar (SOU) 1972:59, Att välja framtid, p. 55.

Hägerstrand, T. 1988. Krafter som format det svenska kulturlandskapet. *Mark och vatten år 2010*. Stockholm, Sweden: Bostadsdepartementet, pp. 16–55.

Hägerstrand, T. 1991. Tillkomsten av nationalparker i Sverige. En idés väg från 'andskap' till landskap. *Svensk Geografisk Årsbok* 191, Vol. 67, pp. 83–96.

Hägerstrand, T., and Godlund, S. 1961. Metod för kommunindelning. Statens Offentliga Utredningar (SOU) 1961:9, pp. 134–147.

Hägerstrand, T., Godlund, S., and Svanström, B. 1967. Samhällsutvecklingen och samhällsplaneringen. Statens Offentliga Utredningar (SOU) 1967:21, bilaga 2, pp. 32–81.

Hägerstrand, T., and Lenntorp, B. 1970. "Tidsanvändning och omgivningsstruktur." In *Urbaniseringen i Sverige: en geografisk analys*. Statens Offentliga Utredningar (SOU) 1970:14. Stockholm, Sweden: Expertgruppen för Regional Utveckling.

Hägerstrand, T., and Lenntorp, B. 1974. Samhällsorganisation i tidsgeografiskt perspektiv. In *Bilagedel 1 till Orter i regional samverkan*. Statens Offentliga Utredningar (SOU) 1974:2. Stockholm, Sweden: Arbetsmarknadsdepartement, pp. 221–232.

Hägerstrand, T., and Öberg, S. 1970. Befolkningsfördelningen och dess förändringar. Statens Offentliga Utredningar (SOU) 1970:14, bilaga 1, pp. 1–55.

Kamiya, H. 1999. Day care services and activity patterns of women in Japan. *Geojournal*, 48, pp. 207–215.

Kamiya, H., Okamoto, K., Arai, Y., and Kawaguchi, T. 1990. A time-geographic analysis of married women's participation in the labor market in Shimosuwa Town, Nagano Prefecture. *Geographical Review of Japan*, Vol. 63, No. 11, pp. 766–783.

Mårtensson, S. 1974. Drag i hushållens levnadsvillkor. In *Bilagedel 1 till Orter i regional samverkan*. Statens Offentliga Utredningar (SOU) 1974:2. Stockholm, Sweden: Arbetsmarknadsdepartement, pp. 233–265.

Mårtensson, S. 1979. *On the formation of biographies in space-time environments*. Meddelanden från Lunds universitets Geografiska institution, Diss. LXXXIV, Lund.

Myrdal, A., Fehrm, M., Frankenhaeuser, M., Hägerstrand, T. Ingelstam, L., Odén, B., Ståhl, I., Engström, A., and Mattsson, Å. 1972. Att välja framtid. Statens Offentliga Utredningar (SOU) 1972:59. In English: To choose a future. Stockholm, Sweden: Svenska institutet.

Nishimura, Y., and Okamoto, K. 2001. Yesterday and today: changes in workers' lives in Toyota City, Japan. In *Japan in the Bluegrass*. P. P. Karam (ed.). Lexington, KY: University Press of Kentucky, pp. 98–122.

Nordström, L. 1938. Lortsverige. Kooperativa förbundets förlag.

Okamoto, K. 1997. Suburbanization of Tokyo and the daily lives of suburban people. In *The Japanese City*. P. P. Kuran and K. Stapleton (eds). Lexington, KY: University Press of Kentucky, pp. 79–105.

Okamoto, K., and Arai, Y. 2019. Time-geography in Japan – its application to urban life. In *Time-Geography in the Global Context*. K. Ellegård (ed.). Abingdon and New York, NY: Routledge.

Öberg, S. 1976. *Methods of describing physical access to supply points*. PhD diss., Royal University of Lund.

Statens Offentliga Utredningar (SOU). 1974. Orter i regional samverkan I. Bilagedel. Grupprapporter. Forskarbidrag. *SOU* 1974:2. Stockholm.

Wang, D., and Chai, Y. 2009. The jobs–housing relationship and commuting in Beijing, China: the legacy of Danwei. *Journal of Transport Geography*, 17, pp. 30–38.

Wang, D., Chai, Y., and Li, F. 2011. Build environment diversities and activity–travel behavior variations in Beijing, China. *Journal of Transport Geography*, 19, pp. 1173–1186.

Wihlborg, E. 2000. *En lösning söker problem*. Linköping Studies in Arts and Science 225, Linköping University. Diss.

Wihlborg, E., and Palm, J. 2008. Pockets of local order for local policy making. *European Spatial Research and Policy*, Vol. 15, No. 1.

5   Transportation and communication research

Spatial movements as part of for daily activity sequences

The individual path and the prism

In line with time-geographical thinking about the indivisible individual, all activities performed by an individual appear in an unbroken sequence and they are performed somewhere. Within this activity sequence, transportation activities are interlaced, thereby connecting the individual's preceding activity at one place with another activity to be performed elsewhere. Hence, all daily travel serve as necessary links between activities performed at different places, and thus they serve as means for the individual to realize a sequence of activities that is meaningful to her. Daily travel, consequently, are put in a wider context.

Because of people's biological constitution, such a daily sequence must contain physiologically necessary activities, like sleep and the intake of food, activities which are largely performed in the home. In the course of the day, an individual may leave the home to perform activities in various projects and afterwards return home for sleep, meals and comfort, which Hägerstrand identified as a principle of return: "This return principle is a constraint, strongly influencing which projects an individual can participate in and for how long" (Hägerstrand 1985: 206).

An individual's movements in the time-space during a day, binding together the activities to be performed at different places, were shown in Figure 4.3 (Chapter 4). To the right in that figure is the activity sequence, or daily activity program,[1] that sets out the individual's time-space movements, which in turn are shown by the individual path to the left. The path illustrates clearly that the individual spends the night at home, and that the home is the point of departure and the point to which the individual returns after completing the activities located at other places.

The prism is a general way to think about and visualize individuals' opportunities to move in the time-space. Time-space moves include stays, since they are movements in time (see Chapter 2). In Figure 5.1 – from Lenntorp (1976) – the prism is related to the theory of distance decay. The upper part of Figure 5.1 shows that the prism is the general basis for the distance decay principle, showing that more journeys are performed in the vicinity of the place where people are located, and consequently there are fewer journeys at a longer distance from a geographic location. The prisms reveal the decreasing frequency of journeys from a place, and the general outcome is shown in the lower part of the figure.

Lenntorp (1976) presents in his PhD thesis the theoretically driven methodological time-geographic research on movements in the time-space. He developed the prism concept and simulation models of people's opportunities to reach various kinds of supply points when traveling by different means of transportation. He used normative daily activity programs to test the opportunities that people with different means of transportation had to fulfill them. Thereby, he underlined the importance of putting daily travel in the context of the activity sequence (daily activity program) and pointed at big differences depending on what means of transportation an individual could use.

The activity-generated transportation research approach was also used to model future travel patterns in a study where the timing of work and services' supply were fixed, based on different assumptions (Ellegård et al. 1977). This study departs from the organization of daily activities in families and a model is developed to show how daily travel is generated by the activities performed. Daily travels, then, are derived from assumptions concerning the scheduling of work and educational activities at the individual and household level. The article concludes that people's needs and daily activities, including scheduling of working time, should be considered in developing the future transport sector, given different assumptions on societal development.

In the development of the prism concept in the late 1960s and early 1970s, Lenntorp (1970, 1976), had to do the very-time-consuming programming manually, since suitable software did not exist. Later, technological developments based

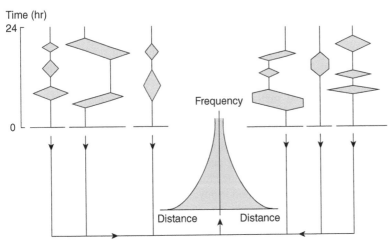

Figure 5.1 Prisms showing the basis of the principle of distance decay (Lenntorp 1976: 63). Reproduced with permission.

on the increasing speed, power and memory capacity of computers led to considerable progress in time-geographical studies. The achievements in GIS (geographic information systems) and GPS (Global Positioning System) were important for the further development of time-geographically-inspired transportation research. However, there is a risk that GPS technologies lead to transportation research that, again, is just concerned with travel activities, while the activities performed while staying at a place, which are the drivers of travel, are left out.

Starting in an article from 1991, Miller has pursued a long-term interest exploring the prism with advanced computer and GIS methods. He elaborated on opportunities to further develop the prism concept theoretically and analytically based on novel technologies, arguing that it would bring new insights to transportation research. The derivation and application of GIS-based prism constructs have been explored and further developments suggested. Later, Miller (2005) developed an analytical theory for measurement of basic entities in time-geography: the individual path, prism, composite path-prisms, stations, bundling and intersections. His time-geographic measurement theory aimed to define time-geographic entities and relationships. He suggested definitions to enable statements about error and uncertainty in time-geographic measurement and analysis and developed and suggested ways for improving opportunities to compare results from different analyses.

Kwan combined GIS and time-geography in the 1990s, and has continuously developed the approach, integrating new data collection and improved visualization methods. She argues (Kwan 2004) that many useful time-geographical constructs were not implemented in analytical methods until the 1990s, and emphasizes the importance of GIS in implementing time-geographic constructs. Kwan (1998) suggests the concept of density surface to display large samples of individual paths in urban areas, showing the daily journeys and stays of people. She worked on this in the 1990s, when the software for such an endeavor was still not available, so she had to do the time-consuming programming herself. She contributed to the analysis of accessibility, measured by the opportunities constrained by the prism. This method can reveal interpersonal differences such as gender, with a focus on the constraints individuals meet in time-space.

Kwan also developed methods for displaying several individual paths in a three-dimensional time-space. Empirically, these studies are based on travel diaries and short questionnaires with a focus on work–home relations, and the individuals' movements clearly reveal the rhythm of urban and suburban areas. She shows that differences regarding gender and ethnicity impact transportation in cities, which should be of importance for urban and regional planners. In Kwan (1999) a network-based GIS method is developed to study space-time accessibility of men and women in European American households in a county in Ohio. Contrary to common ideas, the result shows that full-time working women with a high level of fixity constraints actually travel for longer than men. This conclusion, then, puts into question the dominant idea that the length of the journey to work is a result of low level of fixity. It is obvious that women's fixity constraints are higher than men's, and that increasing female participation in the labor market in itself does not change ingrained gender roles and household division of labor. Kwan has continued to develop time-geographic methods for visualizing activity patterns and in Kwan (2000) a three-dimensional geo-visualization method for this purpose is presented. Empirically it is based on travel diary data from the Portland metropolitan area and these methods are a starting point for the creation of more realistic computational models.

Yu et al. (2008) present a three-dimensional, spatio-temporal GIS model that combines virtual communication with the prism concept. They suggest an adjustment of the prism concept, aiming to identify potential activity opportunities in virtual space from the location of communication channels in physical space, thereby accommodating the study of potential human activities and interactions in both physical and virtual spaces.

Farber et al. (2013) present a method of prism analysis to discover individuals' interaction potential over time in a geographic area. The result reveals the potential time-space in which people can meet in a metropolitan region given the conditions of the prism. The method is based on the time-geographic concept of joint accessibility. It shows when and where individuals' after-work prisms intersect, and can be used in urban planning. Research in a similar direction concerns people's access to services' supply in cities. For example, Widener et al. (2015) analyzed people living in a district with limited supply of healthy food from an activity-based transportation aspect. Earlier research focused on the supply of healthy food near people's homes and concluded that they were therefore determined to purchase unhealthy food. Widener et al. show that people who commute by public transit have a potential to reach supply points to purchase healthy food. By using prism analysis they show that if a trip from work is made by public transportation, if it includes a change of lines, and if there are stores selling healthy food located in the vicinity of the station, then the potential to reach supply points for buying healthy food increases. The waiting time in transit then serves as a facilitator for purchasing healthy food. The findings are visualized with prisms that intersect at the bus stops and train stations when people change line on their commuting trips.

From travel diaries to GPS

Related to mobility and transport, Carlstein (1982) researched livelihoods among nomadic populations. Their activities and movements were studied from both a short (day and week) and long (year, decade, century) time perspective. He shows that the activities depend on the opportunities for sustenance, which in turn are based on the opportunities to use land for grazing and for growing crops. Carlstein puts to the fore two packing principles, one with land that is spread over a large area but used once per year, and the other with a small land area used intensively during the whole year (see Figure 5.2). The differences in the intensity of use in either space or time raise ecological questions from a time-geographic perspective.

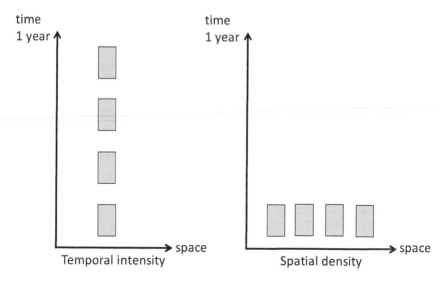

Figure 5.2 Two packing principles for use with time and space: temporal intensity and spatial density (inspired by Carlstein 1982: 152).

Friberg (1990) used time-geographic diaries to collect data about the daily movements included in the activity programs of working women in Southern Sweden. From a gender perspective she combined time-geography with Höjrup's life form theory, and identified specific female types within these life forms. Her conclusion is that women usually adjust their activities, including travel, to the needs of others. This result relates to Kwan's (1999) above-mentioned findings on gender and space-time fixity.

Research on women's opportunities to reach workplaces within a daily prism indicates a political potential of time-geography, not least for identifying and analyzing inequalities that can be approached by policy measures. Scholten et al. (2012) point to the usefulness of time-geography for understanding the conditions of women's daily life from studies of commuting, and analyze the daily struggles to solve coupling problems in the era of mobility from a gender perspective. The authors combine time-geographic tools with social theory, and show that the time-geographic interventional approach reveals obstacles and constraints set up by time-space conditions in women's daily lives. Once attention is paid to such problems it is possible to take actions to improve the conditions.

Frändberg (2008) studied transnational mobility among young people, mainly based on interviews, utilizing the time-geographic individual path as a tool. Based on biographical information and the interviews, the transnational mobility of the young was analyzed from a qualitative perspective. She also investigated the relationship between migration and temporary mobility in relating travel behavior to experiences and life-course transitions.

Tillberg (2001) investigated travel in families generated by children's leisure activities in communities of different sizes. She uses time-geographic concepts to study prisms of the families and found that parents' after-work prisms were heavily influenced and limited by the timing and location of the children's leisure activities. She uses the time-geographical constraints to investigate the outcomes of the social norm that it is good to have leisure activities for children and the professionalization of these activities. Both the timing of meals in the home and the opportunity for the whole family to eat together are constrained by the leisure activities and the transportation connected to them. Therefore, children's participation in organized leisure activities impacts the activity pattern for the whole family to a large extent.

The studies mentioned above did not use information and communication technology (ICT) devices for collecting data. The use of ICT devices, including GPS, for tracking people's daily mobility, visualized with a variety of GIS-based tools, has facilitated data collection for time-geographically-inspired studies on travel and communication.

ICT-based technology and data-driven time-geography

Evidently, each person has to be located somewhere, at a place in the material world. The ICT enables people to communicate without movement in the material world, but their bodies are still located somewhere. ICT, then, allows individuals to follow what goes on at a place where they are not located themselves. In combination with other methods, GPS tools might be powerful for studying everyday life from new dimensions.

In a study from 2008, Kwan combines quantitative and qualitative methods (GPS tracking, GIS representation and narratives) to study how the daily mobility of a Muslim woman is constrained by fears of violence after the terrorist attacks of 11 September 2001. She visualizes this Muslim woman's individual paths before and after 11 September, and the oral history of her days reveals the emotional geographies emanating from fear of anti-Muslim violence.

Birenboim (2016) uses individual tracking devices to identify the time-space movement and experiences of visitors at a festival. A framework is developed that includes the novel concept of momentary experience, which is used for analyzing students' sense of crowdedness and security during the festival. Based on data collected by an app that uses the experience-sampling method, poorly illuminated areas were shown to be experienced as less secure. Shoval et al. (2015) develop sequence analysis methods to show how tourists explore a city during their visit. The researchers define tourist types empirically by applying a sequence alignment method based on the daily sequence of movements of tourists in Hong Kong collected by GPS. This method might be useful in producing destination information for visitors.

One of the first studies to follow the movements of animals was by Baer and Butler (2000), who studied the movements of grizzly bears and used time-geographic concepts. They use the notation system to visualize the tracking done and discuss the opportunities given by time-geography to merge concepts from human and physical geography. Long et al. (2015) used GPS data to develop time-geography-based GIS methods for investigating animal movements, revealing where interactions might appear. The GPS tracks were transformed into time-geographic maps to identify spatial-temporal overlaps. The method is based on the prism concept and joint potential path areas are suggested and tested.

ICT has had a major influence both on developing time-geographical methods and measures and on devices for collecting sequential data. It also influences people's everyday lives. Mobile phones, tablets, laptops and small electronic measurement devices change people's daily activity performance. For example, not long ago, adolescents calling friends were in contact with their friends' family members since anyone could answer the single family telephone, but nowadays every family member has his or her own device, bringing it to wherever he or she is. Thus, the event of answering the phone is performed solely by its owner, and parents do not get any sense of their children's friends by taking the phone call.

In the beginning of the 21st century there were studies on how new ICTs influence people's daily mobility and their everyday activities. Dijst (2013) discusses the state of GIScience (geographic information systems science), where the use of GPS is crucial in the integration of space-time with geography and GIScience, including GPS technologies, human mobility and time.

In a longitudinal study on young Swedes, Thulin and Vilhelmson (2012) use time-geographic diaries and interviews to investigate the effect of ICT in social and spatial contexts. They identify four types of ICT use, called mobility practices (four combinations of what they call *virtual* and *physical* mobility), among young Swedes. Three of the four types are heavy users of the internet and/or mobile phones, while one type rarely use ICTs. They show, for example, that young people with heavy daily ICT communication and large social networks travel a lot. The result indicates that ICT does not replace physical transportation, at least not in a straightforward way, and it shows that the early idea that ICT use would lead to less travel did not come true.

Setting the scene for a number of contributions to a special issue of *Environment and Planning B*, Dijst et al. (2009) discuss changes in people's shopping activities caused by increased use of e-shopping. ICT devices drive e-shopping, wherein a great variety of products can be browsed before purchase. The authors encourage reflection over the effects of e-shopping on the physical realm.

Miller (2005) suggests a people-centered measurement to complement and enhance traditional place-based measurements, in order to better capture modern individuals' activity performance in space and time. It could prepare researchers for investigating activity patterns in a society soaked in ICT.

Shaw and Yu (2009) argue that, since information and communication technologies change human activity and travel patterns, the time-geographic approach should be extended so it can handle the virtual world as well as the physical in a hybrid physical–virtual space. It may have significant implications on everyday life and the human use of space, and offer an analytical environment to study modern society. Yin et al. (2011) suggest an extended time-geographic analytical framework to illustrate how face-to-face meeting opportunities are impacted by access to phones.

Ethical considerations in data collection

Hägerstrand identified problems in collecting empirical data for time-geographic studies about individuals' movements in time-space. He believed that the time-geographical approach should not be limited to empirically grounded studies, but that it ought to be possible to generate valid results at a mere theoretical level (Hägerstrand 1974). However, empirical evidence is essential for credibility in the social sciences.

Since the beginning of the 21st century there has been rapid development in opportunities to gather data on human behavior, from tracks left by mobile phones, computers' internet connections, internet-based payments and the frequent use of various kinds of smart cards. Huge sets of data (big data) of these kinds are, though, not easily available due to ethical rules and commercial interests. In spite of the many new technologies for tracking movements, little research has been presented within geography, as noted by Shoval et al. (2014). Based on a meta-analysis of scientific journal articles, they show that geographers have underutilized the dramatic growth in various tracking technologies.

In Estonia, however, researchers use mobile phones to estimate people's movements in time-space. In a large-scale study, Ahas and Mark (2005) performed one of the first geographic investigations on the use of mobile phone positioning to investigate time-space behavior. Their result suggests a social positioning method that, due to the wide spread of mobile devices, might be used in planning, and they raise some issues on privacy. A small-scale study based on tracking mobile phone data from the Tallin metropolitan area was done by Zhang et al. (2014). It results in an interactive visualization interface displaying data on spatial, temporal and socioeconomic characteristics, and land use.

Schwanen (2016) highlights the risks of utilizing automatically generated big data sets in social science research without thorough reflection, and outlines arguments concerning both new opportunities and risks related to the use of big data in transportation research. He underlines the risk of privileging generality over particularity and calls for deeper reflection, especially when it comes to what situatedness and variations in transportation opportunities imply on a global scale.

The time-geographically-inspired transportation and communication research shows that the individual path and prism concepts are generally applicable and have an important role in research on the horizontally linked, long-distance society, even after the introduction of ICTs. However, sometimes time-geography is questioned as a consequence of the increasing influence of such ICTs. But the foundation of time-geography, the indivisible individual that has to be located somewhere all through a lifetime, is valid even if the opportunities to communicate over long distances and to be in contact with many people at the same time have increased. With a time-geographic take on increased e-shopping, the relevance of the approach for investigating the outcomes of transportation volumes can be seen when considering the numbers of trips for all people who go to pick up packages. Such a time-geographic analysis also urges going beyond the mere geographical movements and recognizing what effects there are on building infrastructure and organization of the retail sector.

E-shopping might influence people's travel activities via the decreasing number of other shopping errands and increasing number of pick-up errands. Usually, people shop for more things at a time while in a physical store, but when picking up e-shopping parcels there is one trip to pick up one package. Then, e-shopping might result in increased traveling. E-shopping may also reduce the number of material retail shops and fertilize the establishment of new physical pick-up stations.[2] This is a continuation of the vertically linked, long-distance society, resulting in a new logistic pattern in the sequence from producer to consumer. Maybe the houses people live in will also change. For example, buildings as pockets of local order in today's Sweden are physically ordered with apartments, which are private spaces that serve as pockets of local order at a micro level. In addition, there are building-specific infrastructures (electricity, water, letter boxes, stairs, elevators and some space for storing things, such as attics and cellars). If e-shopping of food continues to increase, a need will grow for a new material infrastructure in the buildings where food can be delivered and kept cool. The e-shopping might result in huge material investments in buildings and in increased transportation. Taken together, the paradoxical conclusion is that while ICT is claimed to reduce travel, it might lead to increased transportation. A time-geographical, wider analysis of this is urgent in order to avoid adjusting the society to an even more transport-intensive mode.

When involved in the national investigations on urban and regional planning, Hägerstrand wrote about how to create "regional settlements" out of groups of smaller ones, and how the need for transportation in daily life is underlined and problematized. He is clear in his statement: "Transportation is nothing to optimize in itself. Transportation is part of the activities people are engaged in when not on the move" (Hägerstrand 1972: 172, my translation). With these words in mind, it is interesting to see how important time-geography is for the development of transportation and communication research.

Notes

1 In early time-geographic research the concept of the daily activity program was frequently used, and indicated normative programs which were not empirically underpinned. Later the *activity sequence* concept became more common. In this text the two concepts are used in parallel. However, there is a difference in that a program might be regarded as something definitive and very hard to change, while *sequence* indicates that there is some flexibility. The individual consequently constructs her activity sequence via her performing activities, one after the other, and, as the prism shows, there is an opportunity space within which she can be flexible.
2 A little like the old-fashioned post offices . . .

References

Ahas, R., and Mark, Ü. 2005. Location-based services – new challenges for planning and public administration? *Futures*, Vol. 37, No. 6, pp. 547–561.

Baer, L., and Butler, D. 2000. Space-time modeling of grizzly bears. *The Geographical Review*, Vol. 90, No. 2, pp. 206–121.

Birenboim, A. 2016. New approaches to the study of tourist experiences in time and space. *Tourism Geographies*, Vol. 18, No. 1, pp. 9–17.

Carlstein, T. 1982. *Time, resources, society and ecology. On the capacities for human interaction in space and time in preindustrial societies*. Lund Studies in Geography, Series B Human Geography, No. 49.

Dijst, M. 2013. Space–time integration in a dynamic urbanizing world: current status and future prospects in geography and GIScience. *Annals of the Association of American Geographers*, Vol. 103, No. 5, pp. 1058–1061.

Dijst, M., Kwan, M.-P., and Schwanen, T. 2009. Decomposing, transforming, and contextualising (e)-shopping. *Environment and Planning B: Planning and Design*, 36, pp. 195–203.

Ellegård, K., Lenntorp, B., and Hägerstrand, T. 1977. Activity organization and the generation of daily travel: two future alternatives. *Economic Geography*, Vol. 53, No. 2, pp. 126–152.

Farber, S., Neutens, T., Miller, H.J., and Li, X. 2013. The social interaction potential of metropolitan regions: a time-geographic measurement approach using joint accessibility. *Annals of the Association of American Geographers*, Vol. 103, No. 3, pp. 483–504.

Frändberg, L. 2008. Paths in transnational time-space: representing mobility biographies of young Swedes. *Geografiska Annaler: Series B, Human Geography*, Vol. 90, No. 1, pp. 17–28.

Friberg, T. 1990. *Kvinnors vardag. Om kvinnors arbete och liv. Anpassningsstrategier i tid och rum*. Meddelanden från Lunds Universitets geografiska institutioner, avhandlingar No 109.

Hägerstrand, T. 1974. Tidsgeografisk beskrivning. Syfte och postulat. *Svensk Geografisk Årsbok*, pp. 87–94.

Hägerstrand, T. 1985. Time-geography: focus on the corporeality of man, society, and environment. In *The Science and Praxis of Complexity*. Tokyo, Japan: United Nations University, pp. 193–216.

Kwan, M.-P. 1998. Space-time and integral measures of individual accessibility: a comparative analysis using a point-based framework. *Geographical Analysis*, Vol. 30, No. 3, pp. 191–216.

Kwan, M.-P. 1999. Gender, the home-work link, and space-time patterns of nonemployment activities. *Economic Geography*, Vol. 75, No. 4, pp. 370–394.

Kwan, M.-P. 2000. Interactive geovisualization of activity-travel patterns using three-dimensional geographical information systems: a methodological exploration with a large data set. *Transportation Research Part C: Emerging Technologies*, 8, pp. 185–203.

Kwan, M.-P. 2004. GIS methods in time-geographic research: geocomputation and geovisualization of human activity patterns. *Geografiska Annaler: Series B, Human Geography*, Vol. 86, No. 4, pp. 267–280.

Kwan, M.-P. 2008. From oral histories to visual narratives: re-presenting the post-September 11 experiences of the Muslim women in the USA. *Social & Cultural Geography*, Vol. 9, No. 6, pp. 653–669.

Kwan, M.-P., and Schwanen, T. (eds). 2016. Special issue: mobility. *Annals of the American Association of Geographers*, Vol. 106, No. 2.

Lenntorp, B. 1970. PESASP - en modell för beräkning av alternativa banor. Institutionen för kulturgeografi och ekonomisk geografi vid Lunds universitet. *Serie: Urbaniseringsprocessen*, 38.

Lenntorp, B. 1976. *Paths in space-time environments. A time-geographic study of movement possibilities of individuals*. Meddelanden från Lunds universitets Geografiska institutioner. Diss. LXXVII.

Long, J., Webb, S., Nelson, T., and Gee, K. 2015. Mapping areas of spatial-temporal overlap from wildlife tracking data. *Movement Ecology*, 3, p. 38. doi:10.1186/s40462-015-0064-3.

Miller, H.J. 2005. A measurement theory for time geography. *Geographical Analysis*, Vol. 37, No. 1, pp. 17–45.

Scholten, K., Friberg, T., and Sandén, A. 2012. Re-reading time-geography from a gender perspective: examples from gendered mobility. *Tijdschrift voor economische en sociale geografie*, Vol. 103, No. 5, pp. 584–600.

Schwanen, T. 2016. Geographies of transport II: reconciling the general and the particular. *Progress in Human Geography*, Vol. 40, No. 1, pp. 126–137.

Shaw, Shih-Lung (ed.). 2012. Special issue: time geography. *Journal of Transport Geography*, 23.

Shaw, S-H., and Yu, H. 2009. A GIS-based time-geographic approach of studying individual activities and interactions in a hybrid physical-virtual space. *Journal of Transport Geography*, Vol. 17, No. 2, pp. 141–149.

Shoval, N., Kwan, M.-P., Reinau, K., and Harder, H. 2014. The shoemaker's son always goes barefoot: implementations of GPS and other tracking technologies for geographic research. *Geoforum*, 51, pp. 1–5.

Shoval, N., McKercher, B., Birenboim, A., and Ng, E. 2015. The application of a sequence alignment method to the creation of typologies of tourist activity in time and space. *Environment and Planning B: Planning and Design*, Vol. 42, No. 1, pp. 76–94.

Thulin, E., and Vilhelmson, B. 2012. The virtualization of urban young people's mobility practices: a time-geographic typology. *Geografiska Annaler: Series B, Human Geography*, Vol. 94, No. 4, pp. 391–403.

Tillberg, K. 2001. *Barnfamiljers dagliga fritidsresor i bilsamhället: ett tidspussel med geografiska och könsmässiga variationer*. Kulturgeografiska institutionen, Uppsala Universitet. Geografiska regionstudier, 43, 0431-2023. Diss.

Widener, M., Farber, S., Neutens, T., and Horner, M. 2015. Spatio-temporal accessibility to supermarkets using public transit: an interaction potential approach in Cincinnati, Ohio. *Journal of Transport Geography*, 42, pp. 72–83.

Yin, L., Shaw, S.-H., and Yu, H. 2011. Potential effects of ICT on face-to-face meeting opportunities: a GIS-based time-geographic approach. In *Special Issue: Geographic Information Systems for Transportation*. Shih-Lung Shaw (ed.). *Journal of Transport Geography*, Vol. 19, No. 3, pp. 422–433.

Yu, H., and Shaw, S.-H. 2008. Exploring potential human activities in physical and virtual spaces: a spatio-temporal GIS approach. *International Journal of Geographical Information Science*, Vol. 22, No. 4, pp. 409–430.

Zhang, Q., Slingsby, A., Dykes, J., Wood, J., Kraak, M.-J., Blok, C., and Ahas, R. 2014. Visual analysis design to support research into movement and use of space in Tallinn: a case study. In *Special Issue: GeoSpatial*. Gennady Andrienko, Natalia Andrienko, Jason Dykes, Menno-Jan Kraak, and Heidrun Schumann (eds). *Information Visualization*, Vol. 13, No. 3, pp. 187–189. doi:10.1177/1473871613480062. ivi.sagepub.com.

6 Organization of production and work

Work in life and the work life

Two important and interrelated missions set up for time-geography by Hägerstrand (1974) are, first, to link the micro and macro levels without losing important information in the transition between the levels, and, second, to underline the indivisibility of individuals at the chosen level of investigation. This implies a need to increasingly bring the indivisibility of the individual, which ought to be self-evident on the micro level, into investigations of macro-level phenomena. Work might serve as a point of departure for linking the levels and identifying a point of intersection between them. Figures 6.1 and 6.2 illustrate a multidimensional way to look at the intersection between the macro and micro levels, with an indivisible working person as the linking point. In the figures, work is regarded from three perspectives: the production organization (Figure 6.1, upper part), the individual (Figure 6.1, lower part) and the population perspectives respectively (at the bottom of Figure 6.2; compare it to Figure 4.4).

The production perspective of work goes from the national macro level, with its production and service industries (primary, secondary and tertiary industries), via an industrial branch (here, the metal industry) and a company (an automobile firm), into its production units (factories owned by the firm). On the industry and branch level, the statistics on industrial production of a nation may serve as a source. At the company level, information can be collected from the annual reports, while information about production units requires interviews and site visits.

In a lifetime perspective, the person illustrating the individual level works from 16 until 65 years of age. In a year perspective, the individual works for about 11 months, in a week perspective he works for five days and, finally, in the day perspective, for about eight hours. These simplifications give the impression that the time for private life activities increases the shorter the time perspective of a person's life. This is, of course, not true, since the whole working life consists of workdays (eight hours for work) and weekends and holidays with (ideally) no work activities, but when changing timescale some simplifications are necessary. This is an example of how to bring conditions of a short time perspective over to longer time perspectives, having in mind that there is time free

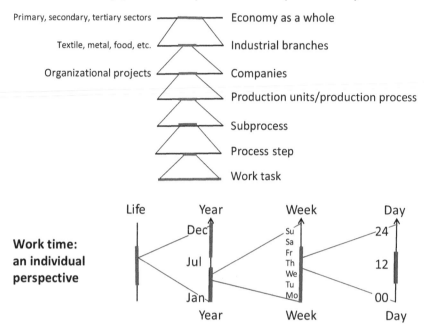

Figure 6.1 Production and individual perspectives of work. Lower part: work activities in an individual's life from the year, week and day perspectives (broad black parts of the individual lines). Upper part: the production perspective from the macro (societal) level to single-work-task (micro) level. (Modified from Ellegård 1983.)

from work when a person is in the labor force. The approach resembles that taken by Hägerstrand to retain important micro-level information about the indivisible individual when talking about phenomena at the macro level, which is indicated in another way in Figure 4.4.

In Figure 6.2 the three perspectives are joined, and the work task as intersection between the work task and the individual is highlighted. Both the individual and the population perspectives focus on time horizons and building on the assumption about the continuously moving now along the Time axis. Time helps show the individuals' involvement in work activities from the whole life (including childhood and old age) to a single day (compare it to Figure 3.10).

Within the chosen production unit, the production process and a sub-process is highlighted. Within the sub-process, process steps may be identified. Finally, the single work task to be performed by a worker is identified. When getting to the levels of the production process, sub-process, process step and work task, it is meaningful to investigate the time needed for performing the work task. Here,

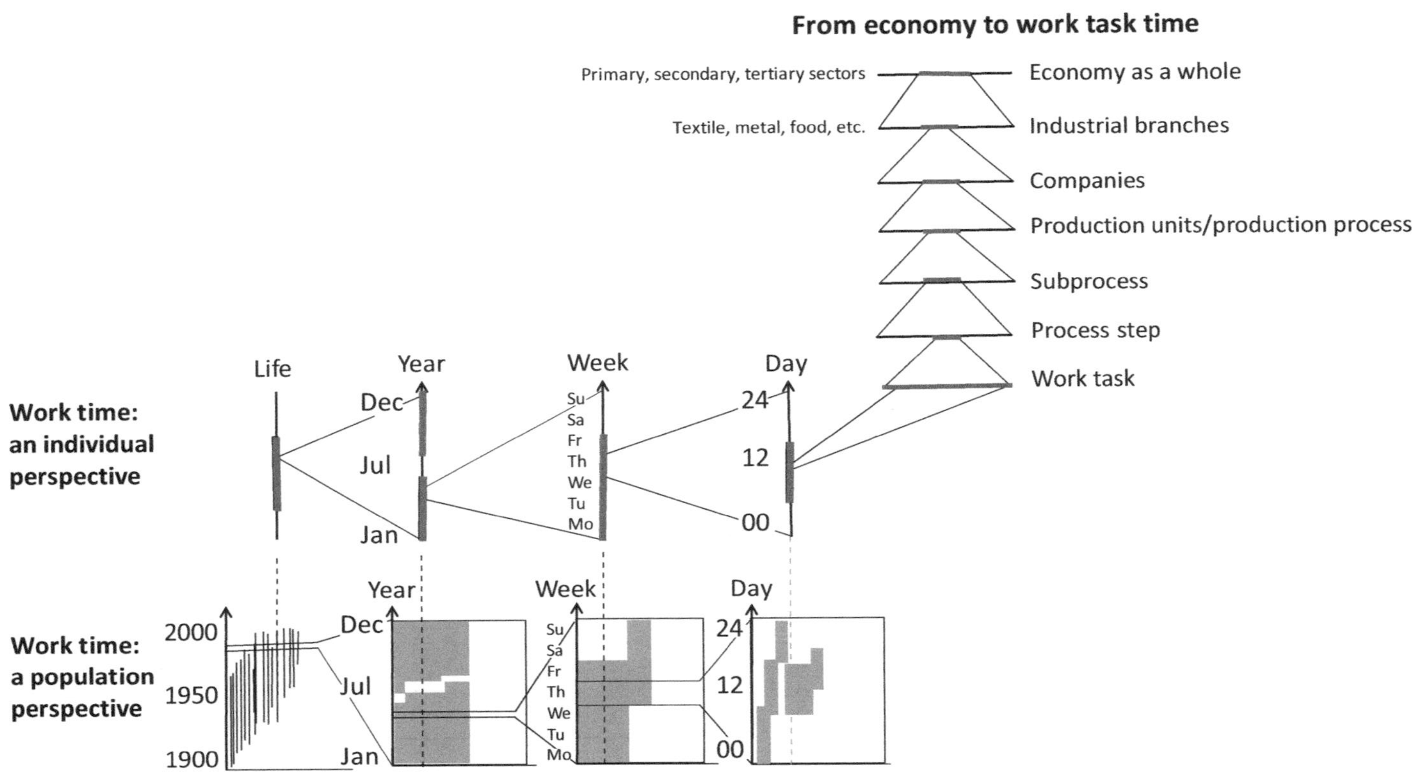
From economy to work task time
Primary, secondary, tertiary sectors
Economy as a whole
Textile, metal, food, etc.
Industrial branches
Companies
Production units/production process
Subprocess
Process step
Work task
Work time:
an individual
perspective
Life
Year
Dec
Jul
Jan
Week
Su
Sa
Fr
Th
We
Tu
Mo
Day
24
12
00
Work time:
a population
perspective
2000
1950
1900
Year
Dec
Jul
Jan
Week
Su
Sa
Fr
Th
We
Tu
Mo
Day
24
12
00

Figure 6.2 The work task as the intersection between production and individual; from macro to micro levels. (Modified from Ellegård 1983.)

the intersection between the production and the individual perspective is obvious. Figures 6.1 and 6.2 are grounded on research on the automobile industry of the 1970s. At that time the duration of each work operation (balanced time) was about one minute. The company organized the whole production process and distributed work tasks over all the employees' work time. Because of the short duration of the work task exemplifying the micro level in figures 6.1 and 6.2, the same work task was repeatedly performed during the whole workday.

The connection between the individual-level day perspective and the work task at the micro level in the production perspective shows for what portion of the day the individual is occupied with performing a single work task. If such a one-minute work task is repeated over the workday and further during the workweek, the workyear and for a large part of working life, there is a risk that the individual is worn out. In Japan, there is a special concept for people getting ill from work: *Karoshi* (National Defense Council for Victims of Karoshi, n.d.). Increasing mechanization, automation and robotization reduce the risk of employees getting worn out, but they also reduce the need for employees in the factories, which is another kind of problem.

In the population time perspective the individual in Figure 6.2 is regarded in the context of all other individuals in the population time of a year, week and day. Since about 50% of the whole population consists of children, adolescents and retired people, about half of the total population time is not used for work activities in the year perspective, more than half in the week perspective and about 25 to 30% in the day perspective. However, much legislation and many rules are adjusted to the labor market and its conditions, which means that work and its time exert strong power not just over employed individuals' working time, but also over the rest of people's lives outside the workplace and over those who are not in the workforce.

Automobile industry

The automobile industry has served as a role model for industrial work organization since the introduction of the assembly line in Ford's factories in the early 20th century (Arnold and Faurote 1915). The MTM (measure-time-movements) system is based on thinking related to the earlier scientific management theory developed by Frederick Taylor. Taylor's management theory aimed to reduce the need for skilled, and thereby costlier, labor (Taylor 1911). For a long time this thinking has structured the education of and thinking among production engineers and managers. Alternatives to the assembly line approach have been presented, and some of them even tried out in real production units, like the Swedish Volvo factories in Kalmar from 1974 and in Uddevalla from 1987 (Sandberg 1995; Shimokawa et al. 1997). In the Kalmar plant the assembly line was literally cut into pieces and work teams were responsible for separate parts of the work on this cut line. In the Uddevalla plant there was no assembly line at all. Instead, a reflective production system was introduced and in use from the mid-1980s until the mid-1990s.

In this section two production systems, a conventional assembly line and reflective production, are presented from a time-geographic perspective, and the discussion focuses on power and opportunities for knowledge development among the employees in the workshops.

Assembly line production

The basic principle behind the assembly line is that objects to be produced (here, car bodies to be assembled into whole cars) are placed on a moving floor, an assembly line, and this line moves through the factory. By the sides of the line are stands and boxes with the material and components which should be assembled on the car body. The assembly line is divided into stations and each station is ascribed a specific time period under which a work task, defined by the manage-ment and production engineers, should be performed by a worker. A work task on the assembly line does not necessarily contain the assemblage of components or materials that are related to each other, regarding neither positioning in the car nor a function in the final product. For example, one part of the same work task could be to set a lamp in place while another could be to affix a carpet. These two parts are connected and put into the same work task just because the time it takes to perform the two adds up to the length of a (one-minute) work task as a whole. The workers have limited overview of work tasks other than their own one-minute task, which makes it difficult for them to see through the logic of the production.

The logic behind the construction of the content of a work task is thereby steered by the time needed for each grip of the hand, not by the internal logic of what materials relate to each other and make up an understandable part of the car. Hence, the knowledge of the assembly logic of the whole car is in the production engineers' department.[1]

The couplings between place (the station at the assembly line and the material stands) and the worker with the right tools and materials are of vital importance for an effective production. In Figure 6.3 there is a principal time-geographic visualization of work tasks on an assembly line. The first part (page 86) shows one worker (visualized with his individual path) doing his single work task on one car body and thereafter he repeats the same work task at the same station on the following car bodies (these are not illustrated separately). The next part (page 87) shows eight workers doing their respective work task on one car body, and the workers are visualized by a section of their individual paths. In the final part, eight workers on eight stations are represented by their individual paths. The many dif-ferent bodies they work on are also illustrated by individual paths.

The path of each worker and the car body he works on move geographi-cally together from the beginning of the station to the end of it. Then they split and the car continues along the assembly line to the next station with the next worker, while the path of the previous worker shows him moving back to the beginning of his station again, to repeat the work task on the next car body. This is repeated all through the day. The individual path for each worker moves in a zigzag-like pattern in the time-space diagram. In the example of Figure 6.3, the eight workers perform their work tasks on 15 different cars (broad sloping lines show the individual paths of the car bodies) during the time span of nine minutes, but each individual worker just does work on 10 car bodies.

Despite its dominance in the automobile industry there are immanent problems related to the assembly line. The assembly line is vulnerable to disturbances, and it does not encourage workers to suggest improvements in the production. Disturbances might appear due to difficulties in assembling a component because of previous misfits, or because a component of inferior quality is delivered to the station. If for such a reason something goes wrong during the performance of a work task, there is seldom time enough within the short duration of the work task for the worker himself to make corrections. Instead the incorrectly assembled car body is left for further assemblage on its journey along the assembly line. Then there are incorrectly assembled parts built into the car and once the car has left the line corrections have to be made. This means that the car has to be partly de-assembled before the correction can be made. Such corrections are very costly and, of course, all production managers do what they can to avoid the need for them.

On the assembly line there are restrictions regarding workers' suggestions to improve the production, even though a worker at any station along the line might have ideas about how to improve the work on his workstation. However, since he has limited overview over what happens before and after the station where he works and no overview over the whole production flow, he cannot make a pre-evaluation of whether the suggestion is suitable for the rest of the line. Therefore, the probability of a worker's suggestion being implemented is limited. It has to pass many production engineers' specialist assessment. There is a need for a big cadre of production engineers to make improvements and to judge whether suggestions from the workers will suit the whole assembly line.

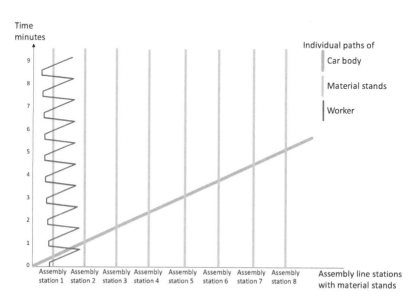

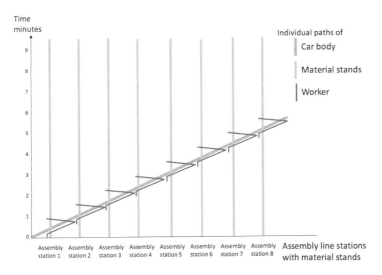

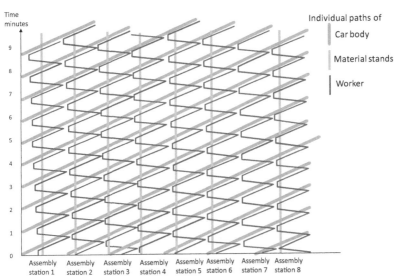

Figure 6.3 Working on the assembly line. First part (p 86): one worker repeats his single
work task. Next part (p 87): eight workers do their separate single work tasks on
one car body. Final part: eight workers do their separate work task, repeatedly
and on many cars.

The management in an assembly line production system has control over the
production flow and it is relatively easy to check if the employees work at the
chosen pace, since this can be revealed by a worker not fulfilling the assigned
work task in time within the geographical borders of his station. It is also possible for the management to increase the pace of the assembly line, since it is

centrally regulated. Hence, for the management there are many power-related pros with the assembly line. However, from a worker perspective, low job satisfaction and wear-out problems in combination with repetitive work tasks are sources of high labor turnover rates and sick leave. This is also a negative for managers, since costs increase for hiring substitute workers and making them learn the work tasks. Factories with such an assembly line production system, however, can employ new workers relatively quickly and put them on the assembly line to perform the simple work tasks. There are still extra costs for such steadily ongoing replacements.

To some extent, the big problems in the Swedish automobile industry in the second half of the 20th century gave rise to efforts to reduce the amount of repetition by introducing so-called job rotation. *Job rotation* implies that a worker performs one short work task for some hours, then he changes to another short work task, and so on in the course of the day. To some extent, job rotation helped improve the wearing-out problems of workers. The effects of tiring and unhealthy work tasks also served as an argument for automation of the heaviest work tasks, especially in the press shops and body shops.

For the top management and the union representatives in the Volvo company, the difficulties of recruiting workers and making them stay were negative effects of the assembly line systems and served as arguments for developing a completely different production system: the reflective production system in the Uddevalla factory (Ellegård 1989, 1997a).

Reflective production

In contrast to organizing automobile production along an assembly line, the reflective production system is based on workers' skills, knowledge and mastering of work tasks with a long duration (Ellegård et al. 1992a, 1992b). A prerequisite is that they know the specific functions and locations of the various components and materials of the car in the assembly process. This production system is different from the assembly line in many respects. The car body is standing still on the workstation all through the assembly process, from being an empty car body to being fully equipped and a complete car. The assembly workers get the components and materials for one specific car at a time delivered on stands, which are specific for that car and on which the materials are ordered so it is easy to find what should be assembled when and where. This so-called "material order" steers the assembly process.

In a reflective production system small teams of four to five workers assemble the complete car together. Consequently, they each assemble 20 to 25% of the total work, and if it takes about nine hours to assemble a car, the duration of their work tasks is about two hours. In Figure 6.4 a workstation with four material stands is illustrated, and the work task performance of two workers during a four-hour period is shown. The car body and the material stands are positioned at the station and the two workers take materials from the material stands and assemble the car.

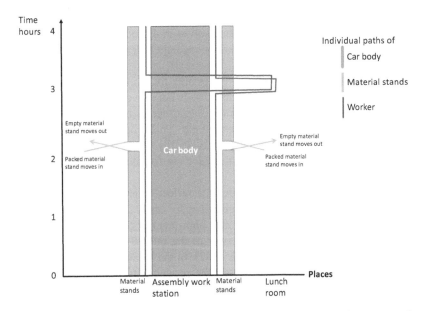

Figure 6.4 Working in the reflective production system. One workstation, two workers, one car body and four material stands during the performance of extended work tasks. The workers, the car body and the material stands are visualized with individual paths of various breadths. The work task lasts for four hours, including lunch break.

Work tasks in a reflective production system are not repetitive since the same work moment is not done more than once or twice per day. The knowledge and competence about car production required by the workers is essential for this production system. The workers gain pride in their work competence. The industrial workers' great production knowledge and the teams' mastering of the assembly of whole cars are grounds for worker-generated improvements in production (Ellegård 2001; 1997b). They control the pace of the work and they know what is happening all through the processing of each car. Thus, their suggestions for improvements in the production are pre-evaluated. They know if a change that they suggest in the beginning of the assembly of a car will cause problems at the end of the assembly of the same car. Therefore, they can pre-evaluate the suggestions they think of and only submit those they find suitable all through the assembly process for further evaluation by production engineers. Such a solution with high competence among the workers implies that the need for production engineers is smaller than in an assembly line factory. In a reflective production system, high worker competence and specialized production engineers is a combination that brings effective utilization of the combined competence to the fore.

This high level of competence and knowledge among workers is, however, also a drawback for the system, since it takes some time to gain the knowledge needed. Therefore, a company leaning on this strategy might be vulnerable to high labor turnover rates. However, if the employees are not physically worn out from repetitive work tasks and if they find themselves in a production system where they can develop their competence and knowledge, the risk of high labor turnover is small.

In a reflective production system, it is possible for workers to correct things that go wrong immediately, since the car body is not moved away from the station. The team of workers themselves give an order to the material-handling department when they are ready for the next car. This production system, then, is driven by the priority given to the correct quality of the products. Contrary to this, an assembly line is driven by the pace set for the line (Ellegård et al. 1992a, 1992b).

In Volvo's Uddevalla plant in Sweden the reflective production system was tried out and the factory ran it on its premises for about a decade. The reason for its ending was a strategic decision at a high level of the company and was not related to the efficiency or productivity of the reflective production system (Sandberg 1995/2007; Shimokawa et al. 1997). In the Uddevalla factory the cars produced were of top quality and so was the productivity in the factory, both resulting from the skilled workforce and the power they had over the pace of production (Berggren 1995/2007).

The time-geographic illustrations of the work tasks in the two production systems in car assembly underline fundamental differences: the individual paths' movements, showing the worker, the car body and the material/components in the time-space, reveal power relations and the degree of repetition in the work. In both cases the workshop and the workstation might be regarded as pockets of local order, but at different scales. If a car factory as a pocket of local order is defined by the production assembly of a whole, complete car, then on one level the pocket of local order of the reflective production system is the team's workstation and it is upheld by the knowledge of the workers, the order of the materials and components on the material stands and the couplings between the workers during the assembly process. On the other hand, the local order of the assembly line is the whole assembly line, for which the order is kept up by the huge physical arrangement of the line in itself, and of the workers' ability to perform the short work tasks assigned to them and ordered by the production-engineering department. Both pockets of local order are created to couple the various components of a car to each other.

Farm work and the dairy industry

The next example of time-geographic studies of work is from food production. The example is taken from a study carried out in Sweden in the mid-1970s and concerns work on a dairy farm and in dairies (Ellegård 1976, 1977; Ellegård and Lenntorp 1980).[2] The idea with this section is to show how some time-geographical concepts are used for analyzing the work of milking cows and dairy production. In the example, the farmhouse and its farmland are regarded as a pocket of local order at one level. The cowshed is a pocket of local order at a more detailed level. The organizational project of the farm is to do farming activities to gain an income for sustenance. Since it is a family-owned farm, the goal of the farming organizational project equals the goal of the household's overall organizational project: to provide a good life for the household's members. The individual project in this example is milking the cows, which can be done by the farmer or somebody else. In summertime the farmer employed a farm hand for milk work.

The order in the cowshed as a pocket of local order was dominated by the temporal routine of milking the cows twice a day, in the early morning and in the afternoon. It was repeated every day, week, month and year. The morning milking activities claimed more work time since at that time the amount of milk was larger. The milking in the mornings was divided into two separate occasions by a break for a breakfast meal. The afternoon milking, which the example will go into more deeply, was uninterrupted. In Figure 6.5 the different time perspectives (week, day and afternoon) are visualized and the performance of the afternoon milking in the cowshed as a pocket of local order in the time-space is described in more detail. The individual path for the milk worker and two aggregate paths for half of the cow herd respectively are illustrated.

The milk worker departs from the afternoon coffee break in the farmhouse and goes to the cowshed, putting the light on, and goes out to fetch the cows on the pasture. Then the cows are put in place and they start eating the hay on the

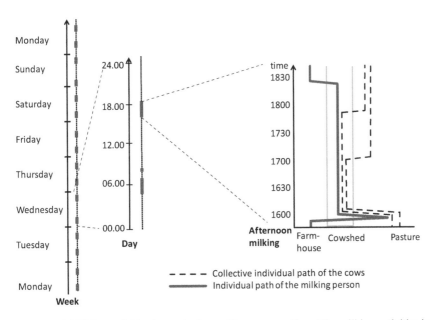

Figure 6.5 Milking activities in week, day and hour perspectives. The milking activities in the project are evenly spread over the week (twice a day). In the day perspective the morning milking is split up by breakfast break, and in the hourly perspective various activities must be performed for fulfilling the work task.

feeding table – which the milk worker put there after the morning milking procedure in preparation for the afternoon milking – and the milking work begins. Once the first half of the herd is milked, these cows are let out of the cowshed while the rest of the cows are milked. When they are ready, they are let out too. Then the milking work task is completed by feeding the calves and tidying up in the cowshed, before leaving it and going over to the farmhouse. The whole work task took about two hours to complete, and, depending on the weather, the health of the cows and the in-calf situation among them it could take a longer or shorter time. The time it took to do the afternoon milking was steered by what must be done rather than by the clock.

Figure 6.6 shows the resources needed for fulfilling the milking work task (there were two milking machines, an extra bottle, a wheelbarrow, electricity and a milk tank), and when these resources are utilized. The milk worker is a resource all the time and so are electricity and the milk container, while other resources are used during shorter periods. The two milking machines are effectively used and no machine is idle during the intensive milking period.

When looking in more detail at the milking work task activities it is clear that the duration for milking each cow differed. Figure 6.7 gives a more detailed picture of how the milk worker's use of the two milk machines relates to specific properties of the individual cows. The properties are indirectly revealed by the

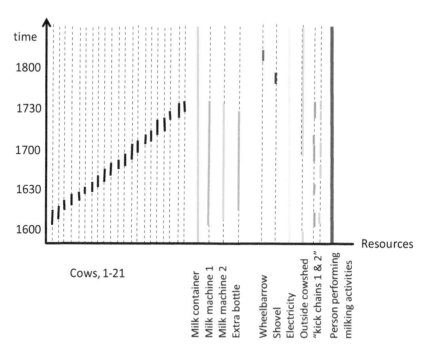

Figure 6.6 Inside the cowshed: 21 cows wait for milking. Resources for performing the activity.

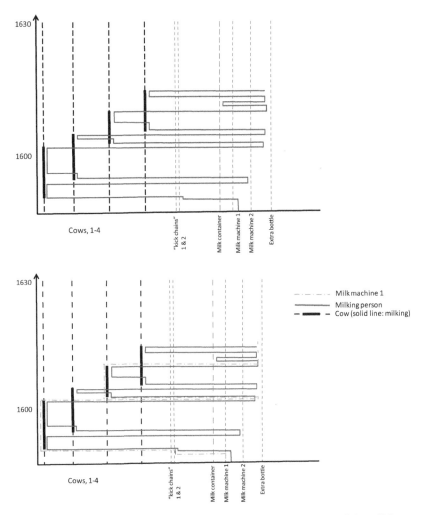

Figure 6.7 Resource use and milking activities. The individual path of the milking
person's movements between the four cows, also indicating the time spent for
milking each cow.

time used by the milk worker beside the cows. For example, cows 1 and 2 are
extremes on the dimension of being easy/hard to milk. Cow 1 needs as much
assistance from the milk worker as possible, since she wants to kick the machine,
while the procedure of milking cow 2 runs smoothly; the milk worker puts the
machine on the cow's udder and takes it off when the machine has done its job.
In the meantime, the milk worker goes back to cow 1 and stays there until she is
ready. Then, the milk is poured into the extra milk bottle and the milk machine
used for cow 1 is now put on the udder of cow 3.

The cowshed as a pocket of local order is vital for performing the activities in the whole farming project wherein milking as a sub-project is the core, since the economic result depends on the sales of milk. However, the overall farming project is much wider, and in order to manage the milking sub-project other, parallel sub-projects must be performed, like growing crops and producing hay to feed the cows, maintaining the fences, growing potatoes and other vegetables for household use, and so on. These sub-projects are not analyzed in this study.

After milking all the cows, the milk tank in the milk room is filled up. This farm has a milk tank with a volume that is adjusted to the number of cows, their milk production capacity and the dairy's milk collecting schedule. Every second day the milk is collected from the farm and transported to the dairy, so the milk must be kept cool in the meantime.

In the vertically linked, short-distance society (see Chapter 1) the milking activities were done at small farms and the production of butter and cheese also took place there. Consequently, no long-distance journeys were necessary. When a dairy industry emerged, more and more dairies were established and the milk was sold from dairy farms to the dairies.

Refining milk into dairy products

Historically milk was mainly produced during the warm part of the year, and in some cases there was a surplus. The milk was then preserved into butter and cheese. In the 19th century, specialized dairies began and they bought milk from farmers and sold the products in markets in urbanizing areas. The English market was of great importance for Swedish farmers in the 19th century in two respects. First, Swedish farmers sold many crops and there was great demand for them in the urbanizing England. However, due to the expansion of exported cheaper crops from North America, which reduced the price of crops, this market fell. The cost of producing crops in Sweden was much higher than in America and the Swedish farmers could not compete. The Swedish authorities were worried about the situation and they supported innovations in dairy industries, since at that time dairy products could not make the long journey from America to England without losing their quality. Second, many farmers were encouraged to produce milk and sell it to dairies, which in turn sold their dairy products to England and later on also to the growing Swedish urban population (Ellegård 1977).

The production chain from milking cows to dairy products consists of a number of inevitable steps. After the milking is done, the milk must be transported to the diary, and once delivered it should be weighed in order to pay the farmer correctly. After weighing the milk is put into the dairy process: first, the skimming and other preparation; second, the parallel sub-processes for the production of cheese, butter and consumption milk respectively; thereafter, packing and transportation to the market. Among these activities, the most dramatic changes relate to the transportation of milk from the farms to the dairies, the skimming technology and the congealing in the cheese-making process. The changes in the ways of transporting the milk to dairies were closely related to developments in other fields: first, of course, the development of means of transportation; second, the development of new technology for stirring the big containers of milk with rennet for congealing; third, new organizational thinking in terms of logistics; and fourth, the introduction of new skimming techniques, which speeded up the process and reduced the space needed for it.

The transportation of milk to the dairies has undergone dramatic changes, both in terms of technology and organization. The localization and number of dairies also influence it. In the early days of the dairy industry, the farmers themselves transported their milk to the dairy, and the yard of the dairy became a social meeting point for the farmers and farm hands who went there once or twice a day. This procedure changed when farmers took the initiative to reorganize the transportation and created teams, which took turns to drive to the dairy with each other's milk. Later on, the bigger dairies bought lorries to go to the farms and collect the milk in bottles, and later still trucks with tank containers. The bigger the dairies were, the more investments they could put into the transportation. However, it was not until it was possible to make cheese in containers with space for more than 100 liters of milk that the motorized transportation of milk to the dairies expanded, followed by a substantial concentration of the dairy industry. In one single county, Skaraborg, in Sweden there were in the mid-1920s over 300 dairies, and the number went down to six in 1976. This concentration process required the motorized transportation of milk, and now the farms had to invest in cooling equipment since the milk was not fetched every day but every second day. The farming project changed as a result of these processes, and some activities are no longer organized and performed by the farmers, such as transporting the milk to the dairy. Since the transportation has been taken over by the dairy organizations, the farmer is decoupled from it. However, new couplings constraints are arising for the farmers, since they have to adjust other activities to the timing of the arrival of the dairy truck. The timing of the milking on the farm in the 1970s was affected by the timetables of the truck since for quality reasons the milk in the farm's milk container had to be cooled down to 4 degrees C before it was collected by the truck. Authority constraints are obvious since the farmer and the milk workers' activities were steered by external organizations' scheduling of them. Thereby, the farmer had to give priority to the dairy organization's timing of the truck's logistics, rather than the natural rhythm of the cows. Big changes have also followed the robotization of milking and the related growth of cow herds on dairy farms. In the 1970s, with semi-manual milking technology, a farm with 20 cows was relatively big, but today dairy farms have at least 100 cows and use robots for milking.

The introduction of new, more resource-effective skimming techniques in the dairy production process during the late 19th century did not affect the farmers as much as the dairies. Initially, milk was skimmed according to the law of gravity; the lighter fat particles of the cream floated slowly upwards in the milk and then the cream could be skimmed off. This procedure took a long time, at least two days, and it also claimed a lot of room. Space was needed for two reasons: first, the containers had to be flat and broad, so that the layer of milk was low and the fat had less milk to float through. Second, cows deliver milk every day and therefore many containers with milk delivered on different days had to stand side by side in the dairies. This spatial problem was partly solved by introducing the "ice method" for the skimming process. This method included milk containers which were surrounded with ice; the cooling made the fat float up faster than the old methods. These two methods, however, had one drawback in common. The milk could not be transported over long distances on the low-quality roads of that time, since the fat particles were crushed into smaller ones, and the smaller the fat particles the slower they would float up. So, when a separator was introduced, based on centrifugal power, many of the dairies' problems were solved: the time it took for skimming was substantially reduced, the distance over which the milk could be transported increased a lot and lorry transportation was encouraged, and the space needed for the skimming process was greatly reduced.

Service professions

The dentist as an assembly line

Production of goods differs in many respects from production of personal services; for example, personal care or treatments. Treatment requires that the provider and the person demanding service get in touch and stay together for the time needed to produce the service, be it hairdressing, medical treatment or dental care. This demand for togetherness is a typical coupling constraint. The service provider cannot treat more than one person at a time, but she can organize the flow of care-takers in such a way that there are no empty time slots in the calendar. A service provider can, if the resources and space in the room for care allow it, set up two care stations and treat one person in one station while the next person waiting for her turn for treatment is at the other station. This kind of organization is used in many modern dental and medical clinics where care is transformed into an issue of efficiency and productivity. The care is organized in a way that resembles the assembly line system of automobile industry. Here the time set aside for each patient is limited according to a normalized standard time for specified types of treatments, and in a way time is given power over the person in need of care.

Figure 6.8 illustrates service supply from the perspective of both the patient and the service provider; it could be a dentist (Hägerstrand 1972). During the period illustrated, the dentist is fully booked with patients and he works without breaks during the whole period (illustrated by the long box). The patients spend some time in the dentist's pocket of local order, including both waiting and treatment, illustrated by the grey parts of the patients' paths. In addition, they have to spend time for transporting themselves to and from the dentist's clinic, illustrated by the black parts of their paths. Thus, the total time spent by the patients in waiting, treatment and journeys considerably exceeds the time that the dentist uses for all the treatments. This way of organizing service production is effective for the dentist, while the patients do not have much of an alternative. The dentist has created a treatment order that resembles that of an industrial assembly line.

All patients in the example of Figure 6.8 use a relatively short time for transportation to and from the clinic. In Chapter 4 an example showed that depending

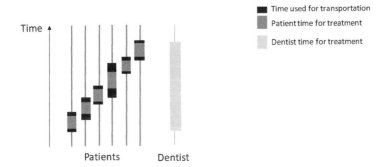

Figure 6.8 Time used by a dentist (service provider) and by patients (service consumers) for treatments during a time period. The patients also use time for transportation to and from the dentist (service provider). (Modified from Hägerstrand 1972: 156.)

on how far from the clinic the patient lives, and whether she has access to a car or uses public transportation, the time that must be set aside for a visit to a doctor (or any other service) can vary considerably (Mårtensson 1974).

Work organization and the work–home relationship

There are studies on work organization leaning on the time-geographic approach. Sometimes time-geographic studies of production and work organization are mistaken for a variant of time-management studies in organizational research, but the time-geographic studies have wider scope and take the indivisible individuals as a point of departure, thereby underlining the influence of work on the conditions for people's overall daily life situation.

Companies and organizations create goals as bases for their organizational projects. This also goes for households as organizations, even though this is in a more informal way. Companies want to make profits from their production and sales of goods and services, while members of households want to create a good life for themselves. As indicated in Figure 6.1 and Chapter 3 (see Figure 3.10), households' and companies' organizational projects meet at the point when an individual performs the work tasks assigned to her. The work tasks are structured and organized so that the company will achieve its goals. The individual who performs such a work task according to the work organization decided upon by the employer has at least one objective – to get paid – and thereby getting the means for maintaining and improving the life of the household. People, however, may also have other goals for their individual work projects than getting paid, like working on things that they like to do or that they feel bring meaning to their efforts.

Trygg (2014; Trygg and Hermelin 2017) investigated the work conditions among employees in advanced knowledge-intensive consultant services. Largely, this kind of work is performed at other places than in the company office; it is a kind of distance work. Here the office as a pocket of local order has a limited role during work days. Some of the work tasks are even expected to be done at home. The studies are empirically grounded on interviews with managers and employees, and time-geographical diaries are written by employees. The daily work conditions of the employees are analyzed by using the time-geographic constraints. The authority constraints reveal that the companies expect the employees to work on the customers' projects with high commitment and long hours. Employees are at the same time expected to fulfill other tasks, such as their own administrative work and preparation, in a flexible way, at home or before and after office hours. This brings severe challenges to their everyday lives and families since the organizational projects of the company and the family collide and cause friction. A general conclusion is that the constraints imposed on the employees in this advanced knowledge industry affect the high labor turnover and rapid reconstruction of companies.

The intersection between the organizational projects of a school and the household of a teacher respectively is of interest since teachers do a lot of work tasks in their homes, just like the qualified knowledge workers of the consultant industry mentioned above. Teachers have a job and an income from it, and in their job they strive to create a good learning environment for the pupils so that they can follow the pupils' learning achievements and experience satisfaction from that. The teachers' core tasks are preparing for lessons, teaching, preparing tests and giving feedback to the pupils about their results and knowledge development. In the Swedish compulsory education system, however, teachers also have to do many other kinds of activities. They are expected to provide care, look after children, discuss them with social authorities and complete administrative tasks. In a study of teachers' work situation in Sweden, Ellegård and Vrotsou (2013) and Ellegård (2014) found that the work week for compulsory schoolteachers followed a rhythm. Monday to Thursday showed similarities, with extended working hours, while on Friday they worked for a bit less time. From Monday to Thursday a considerable portion of the teachers brought work home to do in the evenings; this is visualized in Figure 6.9. Therefore, after dinner and some relaxing they took up work again, doing activities requiring concentration without disturbance from colleagues, school managers or pupils. In this way, many planning, assessment and documentation activities were performed at home in the evening and the overall goal of teachers' household organization, to have a good joint family life, will thereby be counteracted. The teacher as family member withdraws from activities that other family members do, and the schools' organizational project thereby intervenes in the households' organizational project(s). There are many reasons behind bringing work home. The lack of space in the school to perform this kind of task without the risk of being disturbed is one reason. Another is that the time slots between the lectures are too short to do this kind of work that demands concentration in a satisfying way. Still another reason is that teachers are expected to participate in supervision of the pupils during break time.

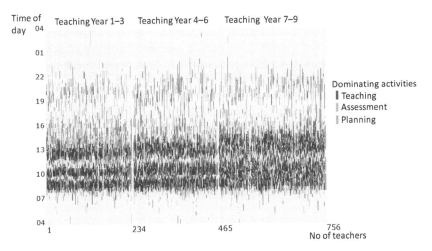

Figure 6.9 Visualization of 756 teachers giving lectures, planning lectures and correcting assignments on Tuesdays. The figure visualizes all teachers' individual activity paths (see Chapter 7), and the periods of each teacher's day when any of the three activities are performed are marked with colors, see legend. The lightest grey is the time when the teachers do other activities. Note the "ribbons" just before 10 and at noon for those teaching years 1 to 6; the breaks are more or less the same in all schools.

When regarding teachers' work at the individual level and from a time-geographic perspective where they are looked upon as indivisibles, it becomes obvious that the schools' organizational idea to give more and more tasks to the teachers does no good for the teachers' workload or their work satisfaction. The research resulted in recommendations such as creating space for calm work areas for teachers in the school, employing administrators to fulfill the administrative tasks that do not require pedagogical skills, and employing assistants for the supervision of pupils during breaks.

In an analysis of work satisfaction as a result of changes in the work organization of a hospital, Bendixen and Ellegård (2014) employed the time-geographical constraint and context concepts. The work life of occupational therapists was investigated by a combination of time-geographic diaries and interviews. The organizational change in the workplace influenced occupational therapists' experience of job satisfaction, but not for the better. Factors constraining job satisfaction were revealed, not least how individuals' personal projects and the organization's project collided. There was a strong engagement from occupational therapists in their patients, but this engagement could not thrive when the new organization did not allow time-space for building trustful couplings to patients. The new work organization resulting from the organizational change in this example did not come out for the best for the individual patient, nor to the satisfaction of the employees.

Nishimura (1998) and Nishimura and Okamoto (2001) investigated the influence of changes in production organization projects – with examples from the Toyota factories in Toyota City, Japan – on the employees' family projects, especially regarding household members' sharing of activities. For families, the work time change resulted in less shared meal time, but also in men's increased performance of household chores. Nishimura and Okamoto (2001) put the living conditions of families with working parents in the context of Japanese work-life culture, and pointed at the restrictions on parents' opportunities to spend time together after factory-shift work hours. The changes are discussed in the context of the transformation of Toyota City from a city dominated by production industries into a modern city and a center for technology based on the automobile industry. This city development was based on close collaboration between the local government and the Toyota company.

Time-geographic research contributes with knowledge of why it is difficult to change an ingrained work organization, and pays attention to the influence of the division of labor in the home on people's work situation (Ellegård 1999). The partition between work and private life is challenged by new technologies, but still work is regulated as a separate sphere divided from the rest of people's lives. Authority constraints, then, have strong repercussions on the rest of people's everyday lives. Time-geographic studies of working people underline the importance of taking the indivisible individual as a point of departure and how her/his activities relate both to work and private life and appear in her/his uninterrupted activity sequence performed in the course of the day. The core point, then, is the indivisible individual performing activities sequentially over time and to analyze when and where work activities are expected to be done within authority constraints, and what coupling constraints they give rise to. Then the effects on people's everyday lives in their household contexts are vital. This is the subject of Chapter 7.

This chapter has presented a time-geographic take on production and work; it is an alternative to conventional approaches where the indivisible individual is not in focus. Hägerstrand reflects in an article on the "good" and the "bad" stress in human life. He discusses what kind of work gives rise to which kind of stress, and wrote:

> I assume that what one would wish everybody is a life in which the happy kind of stress alternates with relaxation. To move in that direction requires an understanding and inventiveness of a kind which differs a lot from the main thrust of today's technology. Perhaps looking at things from new angles could contribute to give our conception of rationality a richer meaning in ways which place people rather the fabricated things at the center of attention.
>
> (Hägerstrand 1984: 18)

Notes

1   In a small car maintenance and repair workshop or garage, the workers, however, need to develop knowledge of how the car as a whole is constructed and what parts and components relate to each other in order to make the car function adequately.
2   The situation has changed a lot since then, both in the farms and the dairies. On the dairy farms milking robots have taken over and the basic structure of the dairy industry is dramatically changed due to small-scale specialist producers of exclusive cheeses and butter.

References

Arnold, H.L., and Faurote, F.L. 1915. *Ford methods and the Ford shops*. Arno, NY: Engineering Magazine Company. Online access: https://archive.org/stream/fordmethods andf00faurgoog/fordmethodsandf00faurgoog_djvu.txt.

Bendixen, H.-J., and Ellegård, K. 2014. Occupational therapists' job satisfaction in a changing hospital organisation: a time-geography-based study. *Work*, Vol. 47, No. 2, pp. 159–171. doi: 10.3233/WOR-121572.

Berggren, C. 1995/2007. The fate of the branch plants – performance versus power. In *Enriching production: perspectives on Volvo's Uddevalla plant as an alternative to lean production*. A. Sandberg (ed.). Avebury: Ashgate Publishing.

Ellegård, K. 1976. Projektet mjölkning, Prövning av begrepp på ett projekt. Mimeo. Februari 1976. Forskargruppen i kulturgeografisk process- och systemanalys. Lunds universitets kulturgeografiska institution.

Ellegård, K. 1977. Utveckling av transportmönster vid förändrad teknik – en tidsgeografisk studie. Mimeo. Forskargruppen i kulturgeografisk process- och systemanalys. Lunds universitets kulturgeografiska institution.

Ellegård, K. 1983. *Människa – Produktion. Tidsbilder av ett produktionssystem*. Meddelanden från Göteborgs universitets Geografiska institutioner, serie B, nr 72, 1983. Diss.

Ellegård, K. 1989. Akrobatik I tidens väv. En dokumentation av projekteringen av Volvos bilfabrik i Uddevalla. *Choros* no. 1989:2. Kulturgeografiska institutionen, Handelshögskolan vid Göteborgs universitet. Göteborg.

Ellegård, K. 1997a. The development of a reflective production system layout in the Volvo Uddevalla car assembly plant. In *Transforming automobile assembly: experience in automation and work organization*. K. Shimokawa, U. Jürgens and T. Fujimoto (eds). Berlin, Germany: Springer-Verlag.

Ellegård, K. 1997b. Worker-generated production improvements in a reflective production system – or Kaizen in a reflective production system. In *Transforming automobile assembly: experience in automation and work organization*. K. Shimokawa, U. Jürgens and T. Fujimoto (eds). Berlin, Germany: Springer-Verlag.

Ellegård, K. 1999. Automation and Inertia. In *Automation in automotive industries: recent developments*. A. Comacchio, G. Volpato and A. Camuffo (eds). Berlin, Germany: Springer.

Ellegård, K. 2001. Redskap för förändringsarbete underifrån – underlag för utveckling av ett IT hjälpmedel. In *Användarperspektivet: strategier för att förstärka samspelet mellan användare och utvecklare*. B. Olsson (ed.). Stockholm, Sweden: Vinnova, pp. 46–54.

Ellegård, K. 2014. Fullt upp från morgon till kväll. *Venue*, 2014.

Ellegård, K., Engström, T., and Nilsson, L. 1992a. *Reforming industrial work: principles and realities in the planning of Volvo's car assembly plant in Uddevalla*. Stockholm, Sweden: Arbetsmiljöfonden.

Ellegård, K., Engström, T., Johansson, B., Medbo, L., and Nilsson, L. 1992b. *Reflektiv produktion. Industriell verksamhet i förändring*. Gothenburg, Sweden: AB Volvo Media.

Ellegård, K., and Lenntorp, B. 1980. Teknisk förändring och produktionsstruktur. En ansats till analys med exempel från mejerinäringen. *Svensk Geografisk Årsbok*, 1980, pp. 75–88.

Ellegård, K., and Vrotsou, K. 2013. En tidsgeografisk studie av strukturen i lärares vardag. *Aktuella analyser*. Stockholm, Sweden: Skolverket.

Hägerstrand, T. 1972. Tätortsgrupper som regionsamhällen. Tillgången till förvärvsarbete och tjänster utanför de större städerna. *Regioner att leva i. Elva forskare om regionalpolitik och välstånd*. En rapport från ERU. Allmänna förlaget.

Hägerstrand, T. 1974. Tidsgeografisk beskrivning. Syfte och postulat. *Svensk Geografisk Årsbok*, 1974, pp. 87–94.

Hägerstrand, T. 1984. Escapes from the cage of routines: observations of human paths, projects and personal scripts. In *Leisure, Tourism and Social Change*. J. Long and R. Hecock (eds). Dunfermline: Dunfermline College of Physical Education, pp. 7–19.

Mårtensson, S. 1974. Drag i hushållens levnadsvillkor. In *Bilagedel 1 till Orter i regional samverkan*. Statens Offentliga Utredningar (SOU) 1974:2. Stockholm: Arbetsmarknadsdepartement, pp. 233–265.

National Defense Council for Victims of Karoshi. n.d. *Karoshi. When the 'corporate warrior' dies*. Tokyo, Japan: Mado-Sha.

Nishimura, Y. 1998. The time-space transformation of automobile manufacturing workers: an analysis based on the concepts of production project and family project. *Japanese Journal of Human Geography*, 50, pp. 232–255.

Nishimura, Y., and Okamoto, K. 2001. Yesterday and today: changes in workers' lives in Toyota City, Japan. In *Japan in the Bluegrass*. P.P. Karam (ed.). Lexington, KY: University Press of Kentucky, pp. 98–122.

Sandberg, Å. (ed.). 1995/2007. *Enriching production: perspectives on Volvo's Uddevalla plant as an alternative to lean production*. Avebury: Ashgate Publishing.

Shimokawa, K., Jürgens, U., and Fujimoto, T. (eds). 1997. *Transforming automobile assembly experience in automation and work organization*. Berlin, Germany: Springer.

Taylor, F. 1911. *The principle of scientific management*. New York, NY: Harper and Brothers.

Trygg, K. 2014. *Arbetets geografi: Kunskapsarbetets organisation och utförande i tidrummet*. Stockholms universitet, Kulturgeografiska institutionen. Diss.

Trygg, K., and Hermelin, B. 2017. Work practice among advanced producer service firms – project work in space-time. *Geografisk Tidskrift – Danish Journal of Geography*, Vol. 117, No. 1, pp. 11–21.

7 Everyday life, wellbeing and household division of labor

Time-geography and everyday life

Introducing time-geography for thinking about daily life

As a consequence of the assumption about the indivisible individual, time-geography underlines the importance of maintaining the sequence of activities performed by people in their daily lives. Empirical time-geographic studies of people's daily lives, then, are based on data about the sequence of activities performed. Aspects like when, where and with whom the activities are performed are important and, depending on the purpose of a study, other aspects[1] can be added. It is vital to relate data concerning such aspects of daily life to each other in order to understand their complexity, and to interpret the couplings between activities, between persons and between persons and resources. Taken together, these aspects provide a starting point to identify the various contexts people are involved in and their internal relationships.

In the 1960s studies about everyday life were performed in the research project *Urbaniseringsprocessen* by Hägerstrand's research group (see Chapter 1). They did observations of people living in communities of different size in southern Sweden and the analysis of this data was aimed at an empirical basis for developing time-geographical concepts. The results were published in research reports (Lenntorp 1970; Carlstein et al. 1970; Mårtensson 1970; Hägerstrand 1970) and in academic PhD theses (Lenntorp 1976; Mårtensson 1979; Carlstein 1982).

The American geographer Allan Pred was a guest researcher in Lund in the 1960s and 1970s, and he was inspired by Hägerstrand's research and made some time-geographically-inspired studies of everyday life. Together with Risa Palm he studied women's couplings to home and children in the modern USA (Palm and Pred 1974). They provide an example of how women's job opportunities depend on the household composition in the cultural context of 1970s America, and used the individual path as a tool. The strength of coupling constraints is underlined by the uneven distribution of labor concerning household chores and childcare, which had consequences for women's opportunities to have free-time activities of their own. This indicated a political potential of time-geography, not least for identifying and analyzing inequalities. Later Pred (1981) published an article where he applied the basic concepts of time-geography in a historic study of how the industrialization process influenced the daily lives of spouses in American families. Miller, another US geographer, published a time-geographically-inspired article about household activity patterns based on historical data (Miller 1982). He studied the introduction of horse carriages and their effects in terms of access to cities for women living in different areas of American cities.

There was some criticism of the early time-geographic studies. For example, Cullen (1972) claimed that the subjective dimensions of individuals' perceptions and experiences were lacking in time-geographic research. He proposed ways to collect data about people's activities combined with data on how they valued the activities. He underlined the importance of considering individuals' activity sequences in line with the time-geographic approach, while at the same time highlighting the need to complement such data with individual subjective data on feelings. This critique was in line with what was later brought up by Buttimer (1976) and others (see Chapter 9).

Perspectives on time in studies of everyday life

In addition to individuals' inevitable anchoring in the time-space, one fundament of time-geography is the indivisible individual. Everybody experiences this on a daily basis in the efforts to fulfill the goals of their many projects, and struggles to act with flexibility when unexpected coupling problems and requirements appear. Time is a fundamental aspect in research on everyday life, but it also brings a problem to research since people experience time very differently. It is problematic to develop a general understanding of daily life and communicate stringently about it, if the discussion is solely based on a subjective, personal, experience of time. Still, this subjective experience dimension is important.

This encourages researchers to think about time from different perspectives. Table 7.1 presents two ways to think about time: the *daily life perspective* and the *instrumental perspective*. Despite their fundamental differences, both are relevant when developing methods for analyzes establishing new knowledge about everyday life. The daily life perspective on time, on one hand, is about how people think about time in performing activities in their daily lives, if thinking about it at all. Time, then, might be important but the activity performed is what really matters. The instrumental perspective on time, on the other hand, is about how time is regarded in research, administration and engineering, where time is measured precisely and can be used as a tool to describe and analyze people's activities and their duration, and also to steer them.[2]

Researching people's everyday lives might be difficult when the daily life perspective on time dominates thinking and communication, not least since each person has her unique interpretation of, for example, when and for how long activities or social gatherings last.[3] However, for a person to recognize herself in a description of her everyday life it must emanate from herself and her experience. If solely building on the daily life perspective on time, problems arise if the

Table 7.1 Perspectives on time (based on Ellegård 1999)

Aspect	Daily life perspective	Instrumental perspective
Use	Used without reflecting, for almost everybody it goes without saying	Used for systematic description, comparison and analysis
Relation between time and activity	Time and activity are closely integrated and very hard to disintegrate	Time and activity are looked upon as different entities and measured by different means
Priority	Activity has priority over the time it takes to perform it (you are ready when your task is completed)	Time has priority over the activity performed (you are ready when the bell rings)
Duration	Focusing 'the present', related to repetition as a central part in forming traditions	Process oriented (past–present–future), linear, though similar events may appear sequentially during a period
Now	'Now' is the timespan that one can control and oversee	'Now' is a steadily coming and disappearing moment; it is the contiguous transformation of future to past
Future	'Future' is the time coming after time one can oversee	'Future' is the time that will be transformed into 'now' in a coming sequence; a prism of possibilities in the future can be revealed from 'now'
Past	'Past' is located in the archive of memory	'Past' is about historical facts that cannot be changed (but re-evaluated)

research aims to get beyond the subjective view and strives to gain a more general understanding of everyday life conditions. Then, a combination of the daily life and instrumental perspectives on time into a *constructive perspective on time* may facilitate comparisons and provide a fair basis for analysis.

A constructive perspective on time implies that data about people's activities should be collected in a way that is grounded in the daily life perspective on time, wherein activity in itself is regarded as superior to the time used, just like it is for the people performing activities. Then, it must be possible to translate the data based on the daily life perspective into concepts and ways of looking at time that recognize aspects of the instrumental perspective on time, like process orientation and a stringent use of concepts related to time.

Based on the daily life perspective on time, the constructive perspective on time recognizes the vital importance of the meaning of the activities performed by a person and thereby recognizes the superiority of activity over the time used for it. Also, the constructive perspective must be aware of the totality of everyday life with its various contexts, relationships and dependencies, when bringing people's own life experiences into the analysis. The starting point is a person's activities as they are performed in sequence during a day (or another time period) and these activities are regarded in the contexts of, for example, who the person is together with, where she is located, how she feels mentally and/or physiologically and physically and what technologies she uses. Thus, the social, geographical, emotional, physiological contexts (and others depending on the purpose of the study) can be related to the activity sequence performed.

Based on the instrumental perspective, the constructive perspective on time should have a process orientation recognizing the flow of time. Thus, it reveals the sequence of activities performed by the person, and the activities can be described and analyzed in the context of other aspects of existence (for example, social, geographical, emotional and technological). It should provide tools that can be used in a similar way irrespective of what person's diary data are investigated. The time-geographical concepts and tools are useful for this purpose.

With the constructive perspective on time as a point of departure, the time-geographic diary method was developed. The constructive perspective inspired both the thinking behind the data collection process (diaries) and the construction of the categorization scheme for analyzing various aspects of individuals' everyday lives. This diary method is briefly described in the next section, and thereafter some studies using the time-geographic diaries are presented. The studies have different purposes and some of them combine the time-geographic diary with other theoretical or methodological approaches.

Diary data from indivisible individuals

In the late 1980s a research project was funded[4] which aimed to develop a time-geographic method to increase the understanding of the doubleness of both diversity and similarity among people in their everyday lives. Each individual had to be regarded as indivisible and, therefore, the conventional methods for studying time use were not suitable.[5] Based on more than 120 open, handwritten diaries collected in the project, an activity categorization scheme and a computer-aided, time-geographically-based method were developed (Ellegård 1993, 1994, 1999; Ellegård and Nordell 1997). The general activity categories and the computerized handling of the diary data facilitated time-geographically-based studies of everyday life's complexity and its rich variety. There are several examples of studies using time-geographic diaries to collect data on activities, togetherness and places visited, of which some are complemented with subjective data on feelings and bodily states.

The time-geographic diary method, based on the constructive perspective on time, prioritizes activity over the time used. This is demonstrated by the diarist herself setting the time when she starts performing an activity. Thereby the activity frames the timing of it, rather than the time framing the activity. With the constructive perspective on time, then, an ongoing activity ends when the next activity starts. This is a smooth way to write the diary, because the diarist just has to look at the time when changing activity. As important for recognition is the fact that the diarist uses her own words for what she is doing. Figure 7.1 exemplifies a time-geographic diary. The aim with the time-geographic diary method is to provide a tool for describing, analyzing and discussing the everyday life of indivisible individuals, based on their self-reported activity sequence, and the results should conform to the diarist's experience of the day. The focus of the time-geographical diary is to gain knowledge about opportunities and constraints at the individual level, which can be used for many purposes; for example, by the individual herself or by care givers or policymakers. However, the diaries can be used for studies at the aggregate level too, which is demonstrated later in this chapter.

In contrast to this thinking, most statistical agencies and others in the time-use research field usually have other aims with collecting diaries, mainly to find out the average time use for various activities of an average individual and to get an overall picture of how time is used in a group or population. Data collection with this purpose is based on diaries in which the diarists are asked to fill in what activity consumed most time within preset time slots, like the Hetus system used by the statistics agencies in the European Union. With such an approach, the individuals are regarded as divisibles. These diaries collect information about the diarists' activity sequence, but when analyzing the diaries the time sequences are collapsed (Hellgren 2015). This approach is grounded in the instrumental perspective on time, thereby letting time slots steer. The time slots have a duration of 10 (sometimes 15 or 30) minutes, and if more than one activity is performed during this time-slot, the diarist is asked to fill in the dominant activity. Obviously

Time	What I do	Where I am	Together whith	Reflection	Comments	....	.....
0600	Wake up	At home	Mate	Tired			
0615	Take a shower		Alone				
0635	Eat breakfast		Mate	I was hungry nice to be served			
0705	Dress		Alone				
0715	Go to the bus stop	On the road	Alone				
0725	Go by bus	In the bus	Many other passengers Bus driver	I'd like to go by bike but the weather is too lousy	Why are there so few buses when they are so crowded?		
..and so on							

Figure 7.1 The headlines given in the time-geographic paper diary. An example of how the diary is written, using the columns of interest for the diarist.

activities with a short duration will be underestimated. Consequently, an everyday life described from such a diary will differ from and contain fewer details than results from time-geographic diaries. This is no problem at a general group level, but it is not valid for conclusions at the individual level. This is elaborated on in Ellegård (2006).

Another difference between the time-geographic diary method and Hetus-like methods regards the categorizations of activities. The activity categorization scheme used for the time-geographic diary method is empirically generated and developed by a bottom-up approach (Ellegård 1994, 1999, 2006). It is based on thousands of diary entries of activities, and some activity categories should not even have been thought of with a top-down approach[6] in constructing activity categorization schemes.

The time-geographic activity category scheme takes its point of departure in the diarists' lives, where the overall project for all people is assumed to be "to live life". Any kind of project can be performed within this common overall one, ranging from lifelong ones, like raising children, to shorter projects like purchasing a new dishwasher. This flexibility is of great importance for understanding everyday life from the perspective of the indivisible individual. Hence, within the overall project all other projects with their diverse goals are formulated and activities are performed by the individual to achieve them.

The time-geographic activity category scheme is hierarchical, with five levels of detail (see Figure 7.2). There are seven categories on the highest (sphere) level[7] (see Table 7.2a). When used for categorizing a diary even the highest-level activity categories give a lot of information about general aspects of an individual's daily life. For example, by distinguishing between care for others, household care and procuring and preparing food, gender differences are revealed. The totality of the activities at all levels in the scheme gives room for a capturing a great variety in everyday daily life.

The category scheme in the Hetus method has six categories on the highest level (see Table 7.2b). In this scheme all kinds of household chores are put into the same category (domestic). This may result in learning, as in the Swedish national time-use survey from 2010–2011, that the gender difference is masked and that women and men spend about the same time for household chores (Statistics Sweden 2012). This is true, but when looking at household chores, the next level down, it becomes clear that men do more maintenance work and household administration than women, while women perform more childcare, housekeeping and food-related activities. When the time-geographic category scheme is used, this kind of difference will appear at the highest level.

The time-geographic diary method is presented in depth in Ellegård and Nordell (1997) and in Ellegård (1999), where the principles behind categorizing and visualizing the diaries are also presented. In Ellegård and Nordell (1997) the physical and mental states of the diarist are also included. It is shown how a diarist uses her own diary as a reflective tool, which, in combination with recurrent discussions with a researcher, result in her making changes in her daily activities to live a more satisfying life.

Overall goal/project: To live life—Living one's life

7 spheres: CARE FOR ONESELF, CARE FOR OTHERS, *HOUSEHOLD CARE,*
REFLECTION/RECREATION, TRANSPORTATION, PROCURE
AND PREPARE FOOD, WORK/EDUCATION

Categories in the household care sphere:
CARE OF HOME, CARE OF CLOTHES, CARE OF THINGS,
HOUSEHOLD ADMINISTRAITON,
SHOPPING, GARDENING, HANDICRAFT

Classes in the category care of home:
tidyup, make the bed, pick upthings, indoor flowers

Sorts in the class tidy up: *cleanup,* thorough cleaning

Specifications in the sort clean up: sweep, vacuum, scrub, dust

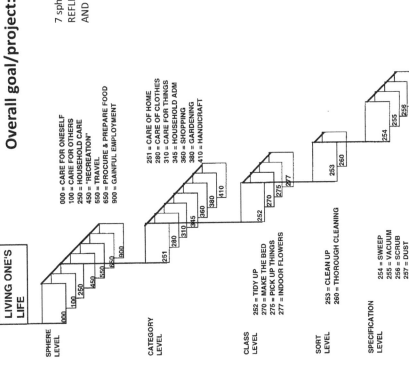

SPHERE LEVEL

LIVING ONE'S LIFE

000 = CARE FOR ONESELF
100 = CARE FOR OTHERS
250 = HOUSEHOLD CARE
450 = "RECREATION"
550 = TRAVEL
650 = PROCURE & PREPARE FOOD
900 = GAINFUL EMPLOYMENT

000 100 250 450 550 650 900

CATEGORY LEVEL

251 = CARE OF HOME
280 = CARE OF CLOTHES
310 = CARE FOR THINGS
345 = HOUSEHOLD ADM
360 = SHOPPING
380 = GARDENING
410 = HANDICRAFT

251 280 310 345 360 380 410

CLASS LEVEL

252 = TIDY UP
270 = MAKE THE BED
275 = PICK UP THINGS
277 = INDOOR FLOWERS

252 270 275 277

SORT LEVEL

253 = CLEAN UP
260 = THOROUGH CLEANING

253 260

SPECIFICATION LEVEL

254 = SWEEP
255 = VACUUM
256 = SCRUB
257 = DUST

254 255 256 257

Figure 7.2 The hierarchical category scheme developed for the time-geographic diary method. The overall goal for everybody is supposed to be "to live life", and there are seven main categories (spheres), which are further broken down to the most detailed (specification) level, as exemplified by the sphere "Household care".

Table 7.2a Highest-level activity categories in the time-geographic activity
categorization scheme

Activity category	Relevant activities
Care for oneself	Eat, sleep, care
Care for others	Help others to eat, sleep and get care
Household care	Care for the home, belongings, administration
Transportation	Movements to and from locations
Reflection/recreation	Get inspiration, relaxation, socializing, playing
Procure and prepare food	Cultivate, shop for food, cook, bake, and so on
Work/education	Work and education

Table 7.2b The Hetus categorization scheme (source: www.h6.scb.se/tus/tus/Default.htm)

Hetus main categories
Employment
Domestic
Study
Personal care
Leisure
Travel

Time-geographic visualizations of contexts in everyday life

When writing time-geographic diaries people start thinking about what they do.
This reflection process is deepened when the diaries are visualized and shown
to them. The visualization is based on the information about: places and movements; activities; companionship; technology use; emotions; and other aspects
of life which are collected depending on the purpose of the study. The visualization is a point of departure for discussion about daily life as it is lived and
about what changes can be done to make it comply better with the wants and
needs of the diarist. Here, seemingly simple issues are fundamental, like when
and for how long an activity goes on and if the activity fragments into smaller
parts, and in what contexts the activities appear. Knowledge about such issues
may help to better understand stress on an individual level, physical exercise
and conditions of everyday life for people with impairments and also household
division of labor.

The visualizations and analysis of time-geographic diaries illustrate the data
with several kinds of paths, and the focus is always on the individual diarist (see
Figure 7.3). There is the time-space individual path (showing the geographical
context) and the activity-oriented individual path, both symbolizing the diarist's
daily life, but with a different focus. The time-space individual path shows the
diarist's movements between places over time (as shown in Chapter 3), while the
activity-oriented individual path visualizes the activity sequence of the individual;

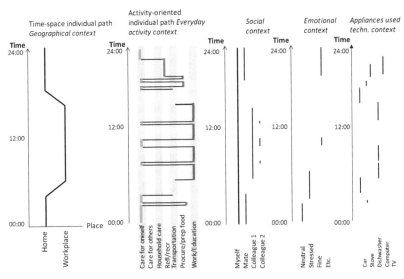

Figure 7.3 Visualization of contexts generated from time-geographic diaries. The geographical context shows the time-space individual path illustrating the places a person has visited and the movements between them. The everyday activity context shows the activity-oriented individual path, illustrating the activity sequence of the person (activities in the least-detailed level with its seven sphere categories). Here the horizontal lines in the activity context indicate change of activity, while transportation is an activity of its own. The social context shows with which other people the diarist has been with in the course of the day. The emotional context shows what the diarist wrote about her feelings during the day. Finally, the technological context shows what appliances the person used in the activity sequence.

hence, it shows the duration of activities and when one activity is followed by another (see figures 2.3 and 3.1).

Social companionship, which constitutes the social context, is visualized by a path over the whole day symbolizing the diarist, since she is there all the time. It is important to draw this path since it makes the *relationship* between the diarist and other people obvious. In addition, there are sections of other individuals' paths visualized during the period of the day; these sections illustrate when the diarist has noted that she is together with these specific people. Of course, from the perspective of the other people, they should be illustrated as full-day individual paths, but in this context it is the section where the companionship is manifested that is of interest.

Sections of paths are also used to visualize the diarist's technologies used during the time periods when the diarist has noted such use: the technological context. The same visualization principle, with sections of paths, is used for emotions and other experiences: the emotional context. The same principle can also be applied when visualizing secondary activities.

The visualizations of the geographical context, the everyday activity context, the social context and the emotional context[8] make it possible to identify the *links between what is happening when in the different contexts*. For example, what activity the diarist performs where, and how she feels emotionally when she is doing this activity together with a certain person, can be traced. In Figure 7.4 this process is exemplified by horizontal help-lines, helping the reader to link the different contexts to each other at these certain time periods. This can provide a better understanding of, for example, what combinations of contexts result in positive or negative feelings.

The visualization also offers opportunities to investigate what activities are performed in order to achieve the goals of different projects an individual engages in, which constitutes the project context. Hence, the activities that the individual performs to achieve the goal of a project are marked. The result illustrates that the activities involved in one project appear here and there in the everyday activity context (equaling the activity sequence) of the diarist. The individual's activity sequence, then, contains activities from different projects and they are mixed in the activity sequence (see Figure 7.5). From such an analysis, the constraints that a person experiences in daily life are put to the fore and can be challenged.

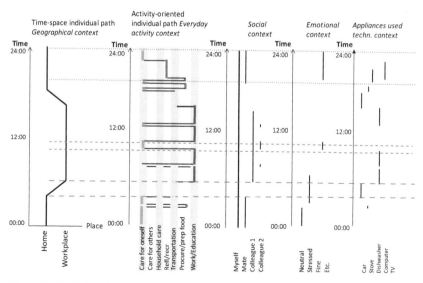

Figure 7.4 Relating the contexts to each other. Via "help-lines" the information given about the various contexts the person was involved in (as shown in Figure 7.3), can help in finding out, for example, where a person was located at a certain time, what she did there, who she was with when performing the activity, what feeling was dominant and finally what appliances she used (if any). At lunchtime, the person illustrated in the figure is at the workplace, performing a lunch activity, together with two colleagues. She feels fine and does not mention any use of appliances.

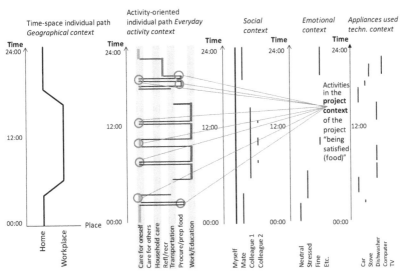

Figure 7.5 Finding the project context in the activity sequence of the everyday activity context. A project is fulfilled by individuals performing activities that appear here and there in the everyday activity context. The figure shows activities that altogether constitute the project context for "being satisfied" in the day perspective. This individual performs activities in the sphere "procure and prepare food" twice during the day and eats five times.

Computer-based collection and handling of time-geographic diaries

Based on data from time-geographic diaries, a computer program[9] was developed to visualize the individual's diaries day by day (Ellegård 1994) and it has subsequently been further developed (Ellegård and Nordell 2011). The computerized visualization shows several contexts in which the individual is involved during a day: the activity context, the geographical context, the social context and the emotional context. The software is flexible and can be adjusted for the purpose of various studies.[10]

Since the turn of the century, cell phones have become widespread and there is an increasing demand for diary collection by apps.[11] Based on the time-geographic diary method and its activity categories, Vrotsou et al. (2014) created a smart phone app, PODD, for electronic collection of diaries. The diarist carries the PODD diary and makes notes in real time.

Thus, the diarist does not need to think about time, just about putting in information like *activity, secondary activity, place, emotions* and other things, which are asked for depending on the purpose of the study. The data are sent to a server and are used for producing activity sequences and visualizing them for further analyses at both the individual and aggregate levels. The software produced for handling the time-geographic diaries is interactive, which implies that various hypotheses and ideas about activities in daily life can be tested. There is another application for collecting individual diary data on cell phones based on the time-geographic diary principles which has been adapted for clinical use of occupational therapists. Initially used for research in the field, and developed by Kroksmark and Nordell, it is called A Day. A computer-aided time-geographic method to investigate the life perspective was developed by Sunnqvist et al. (2013). It aims to construct long-term biographic life charts of individuals. Examples of how time-geographically-inspired diaries are used in research are presented in the next section. The utilization of these programs is shown later in this chapter.

The everyday life of the indivisible individual

There is an increasing number of time-geographically-inspired studies on everyday lives of people, not least people in vulnerable situations. There are studies of this kind in a variety of academic subjects, such as human geography, social work, psychiatry, occupational therapy, physiotherapy, political science and technology and social change. With the indivisible individual and the principle of letting the diarists' own words steer the information in the diaries, the time-geographical diary method has attracted special interest from occupational therapists. At the core of occupational therapy is the drive to facilitate an everyday life with dignity and pride, and for people with difficulties to perform daily activities and thereby achieve the goals of the projects they want to pursue. Hence, time-geography is used to ease the coupling constraints that arise from the capability constraints. Authority constraints may also severely impact the daily lives of people with capability constraints. Time-geographic analyses can be useful to show to authorities that the rules they have set up are dysfunctional.

In general, the time-geographic diary method is used as a tool to find out what is needed to overcome capability (capacity), authority and coupling constraints and how to organize pockets of local order and projects that the individual can master and thereby live life as well as possible. There are also studies on the household division of labor and energy use in the home. Finally, there is a time-geographic method utilizing diaries to visualize aggregate activity patterns, where the indivisible individual is recognizable on the aggregate level. This meets Hägerstrand's (1974) requirement that important information about the individual level should not get lost when presented at aggregate levels.

Individual-level analysis – sustenance, health and wellbeing

There are several studies in occupational therapy, physiotherapy, psychiatry and social work inspired by the time-geographic approach. They concern, for example, stress among women with double responsibilities; elderly men and their health related to their physical activities; how to capture the life story of people with suicidal behavior; women subject to abuse; and how daily life is lived by people with psychiatric diagnoses living in their own apartments with social service assistance. Many of these applications use the time-geographic concepts, methods and visualizations in order to help researchers and service staff to understand the life situation of these people and to communicate with them. Since the communication is based on diary notes that these persons themselves have produced, it is relatively easy for them to recognize themselves and their own daily lives. Visualizations of the person's own daily activity sequence are of vital importance for discussions about their daily life, since they help reveal couplings and phenomena that usually are hard to articulate because they are too self-evident to be thought of. The outcome of such discussions creates a basis for further treatment and assistance. In a seminar on time-geographic research, a researcher, Karin Örmon, cited from her work in which women wrote time-diaries, which were used for discussions about their life situations. One woman said: "Finally, my life story is my own."

Several PhD theses on the everyday life of people in vulnerable situations which in various ways utilize the time-geographical diary method have been written in Scandinavia.

The PhD thesis in human geography by Nordell (2002) investigated how women with body pain changed their daily lives. She combined time-geographic diaries with group treatment programs and discussions. The women in her study were helped by both the combination of discussions and the visualizations of their diaries and they found that their own daily life was well represented in the visualizations. This gave them motivation to realize changes in order to get better control over their health. There was a long-term effect of this combined treatment program.

In her PhD thesis on human geography, Westermark (2003) used time-geographic diaries and reflective diaries[12] in combination to study low-income women's everyday lives and sustenance in urban Colombia. She let two women intermittently write the two types of diaries (time-geographic diaries of the day and reflective diaries) during a four-year period. The time-geographic diaries showed the women's daily struggle to create performable activity sequences for getting enough money to feed their families and, at the same time, to build up long-term projects for their own and other women's sustenance. The reflective diaries covered their whole life stories, which widened the understanding of how authority constraints hampered their efforts to sustain themselves. The diaries were combined with recurrent interviews and from this data the women's projects related to their struggle for sustenance are analyzed. The time-geographic diaries helped acknowledge the problems of these women appearing on a daily basis; for example, regarding when coupling constraints involving their small children impinge upon the projects they have to pursue for their sustenance. The individuals' life courses and daily activity perspectives are utilized to show how institutions can produce development projects that meet the indivisible women's needs.

The PhD thesis on human geography by Kjellman (2003) studies how drug addicts and old people with intellectual disability found and experienced a place when cared for by social authorities. Kjellman combines time-geography, sense-of-place theory and socio-geographic theories of marginalization and a person's prospects to acquire a place, thereby integrating experiences of people in marginalized groups with time-geography. The result puts into question some of the political reforms of the policy to move old people with intellectual disabilities to their own apartments at another place, sometimes even to another town. These people had spent their whole life in institutional care and lacked experience of living a self-controlled everyday life. The reforms did not consider the indivisible individual and the importance of social and material stability in the pocket of local order in which these old people live.

The occupational therapist Ulla Kroksmark played a major role in introducing the time-geographic approach to occupational science. She initiated a Scandinavian network on time-geography for occupational scientists and physiotherapists. One outcome of this network was a special issue of the *Journal of Occupational Science* (2006, Vol. 13, No. 1) wherein several Scandinavian occupational and social science researchers presented their work based on time-geographic diary data collection.

Kroksmark and Nordell (2001) investigated the potential for using the time-geographic approach in occupational science and its practice in occupational therapy in a pilot study comparing how adolescents with and without visual impairment spent their days. Young people with low vision experienced fewer opportunities than young people without visual problems. Adolescence is a sensitive period of life, when young people usually develop independence from their parents, thereby freeing themselves from authority constraints. This article clearly points out difficulties in this process for young people with low vision. They depend on parents for transportation (coupling constraints) and perform fewer activities with friends in their daily activity sequence compared to sighted young people, with much depending on their capability constraints.

In her thesis, the Norwegian occupational therapist Eva Magnus (2009) used time-geographical diaries and interviews to show how the conditions of daily life for university students with limited physical abilities were influenced by their capability constraints. Magnus uses time-geographical diaries and the constraint concepts to explore how students with disabilities use time for activities related to their studies and how they try to overcome their capability constraints. The study showed clearly that most activities, private as well as study-related ones, take more time for a student with impairment, due to their capability constraints. It showed that neither the material nor the social environment were always helpful to these students. Instead, these environments caused severe coupling constraints in everyday life for these students, which added to the extra time they needed for their activities related to their study mates.[13] The results are useful for policymakers who strive to create regulations that support equal opportunities among university students. The new knowledge from Magnus' dissertation influenced national policies in Norway and the conditions for students with physical impairment were improved by allowing them longer study time in combination with changed conditions for their study financing.

In her PhD thesis on technology and social change, Åström (2009) shows the complex influence of coupling constraints when introducing computer aids for schoolchildren with visual impairment. The manifold service providers involved, the school building with its room layout and corridors, the transport arrangements, teachers, low-vision service care providers, computer-aid providers and families are so complex that the outcome makes it difficult for the children to reach their potential in learning. Just getting the computer aids was not enough to achieve the high policy goals set up regarding equal opportunities for all pupils. One reason is that the many different organizations involved take their own perspective respectively as a point of departure rather than the perspective of the indivisible child, and in this case it concerned the necessity of coupling the child with her computer aid when doing school work at school and in the home.

In physiotherapy, the importance of bodily physical activities is key. After retirement, men are at risk of decreased physical activity partly because they are usually not involved in housekeeping activities in their households. Bredland et al. (2015) combined time-geographic diaries with codes for metabolic equivalents (METS) to find out more about the level of physical activity among elderly men. The result points to the importance of everyday physical activity, and says that mundane activities, such as household chores like vacuuming and walks to the store, in combination may provide healthier physical activity levels. For the sake of physical activity, it is not necessary to get involved in programs at the gym. However, there is a social dimension in the gym as a pocket of local order which should not be underestimated. The results could be important for developing activity programs for older men.

The MD thesis in occupational therapy by Erlandsson (2003) regards everyday life and stress among women. The time-geographic approach was inspirational for the visualizations and ways of discussing sequences of occupations (activities). The study shows that more fragmented activity performance increases the stress level. It also shows that together with theories in occupational science, the time-geographical approach helps in understanding the relationship between occupational balance and health among working women.

In the 1990s Lenntorp wrote a piece called "Biographies, diagnosis and prognosis". This was in the background to the work by Sunnqvist (2009) and Sunnqvist et al. (2013), where a time-geographical method was developed in order to understand and prevent suicide attempts. This is the first psychiatric study where time-geography is utilized as as a means to understand the prehistory of suicidal behavior. Based on interviews, a time-geographical life chart is produced in order to investigate the individual's life. Information was collected from people on geographical moves to new places, social and burdensome events before their suicide attempt, and other important events, like the birth of siblings, marriage and divorce. The life chart diagrams are used as background for discussion with the people about suicidal behavior and the events influencing it (Figure 7.6). The events in the life chart are related to the person's coping strategies. Thus, the pathway to suicidal behavior is illuminated and may be used in future to prevent suicide attempts.

In her PhD thesis in social work, Andersson (2009) investigates the daily lives of people living with psychiatric diagnoses. She introduced time-geographical concepts in analyzing people's everyday life projects, depending on whether the individual or the social institution has power over these projects. She shows the

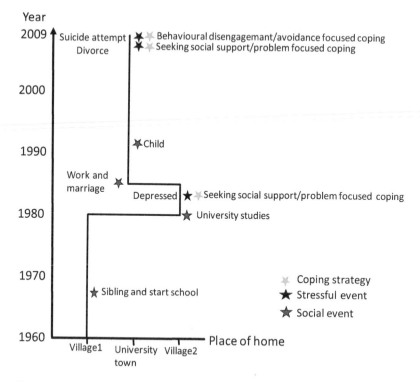

Figure 7.6 Illustration of the principle for creating a life chart, used in research on suicide attempts. From interviews important life events are identified. The person's coping strategies for stressful events and their suicide attempt are included in the chart. (Modified after Sunnqvist et al. 2013: 340.)

role of the social worker providing "living at home support" and assisting the person to perform basic daily activities. The importance of taking the indivisible individual into consideration when organizing social support for people with psychiatric diagnoses is underlined. Andersson gives thought-provoking examples of what difficulties appear for the people if the institutional routine arrangement fails. The secure, daily routines are broken for the individual and the resulting situation can be hard to handle.

The human geographer Martin Dijst (2014) investigates stress in different urban settings. He integrates stress with the meaning of modernization and urbanization in two societies, one individualistic (the Netherlands) and one more collective (China), and presents strategies for handling relationships with social environments and regulating stress and wellbeing.

The studies presented above mainly relate to the individual level, even if there are, of course, social relationships which influence the individual in her performance of daily activities. There are also some time-geographical studies on everyday life at the household level.

Household-level analysis

If time-geographic diaries are collected from more than one individual in the same household, the division of labor in the household can be thoroughly researched, providing a basis for household-level analyses.

Orban (2013) investigated how to assist parents with obese children to change their activity pattern in order to avoid child obesity. The aim was to see if a one-year intervention, with time-geographic diaries and reflective discussions on daily activities, made the parents spend more time doing more healthy activities with their children. Different patterns of parents' shared daily occupations were identified from the time-geographic diaries, separately written by both parents in the same families. The visual representation of the diaries facilitated parents' reflections on their daily activities, and the discussions with the occupational therapist focused on the effects on the obese children. A follow-up study showed that the body mass index of the children as a group had decreased. Orban et al. (2012) investigate how the time-geographic diaries facilitate reflections on change in everyday life.

In a study of consumption in industrial workers' households, Ellegård et al. (1993) used time-geographic diaries and added a column where the direct cost caused by the activity was noted. Both spouses wrote this kind of time-geographic diary over a whole week. The research aimed to make the participants aware of the total sum of money that the small but frequent purchases they made added up to over the week. The result showed that they yielded better control over their finances through their engagement in the project.

In a different empirical field, energy use in the household sector, Isaksson and Ellegård (2015) demonstrate the importance of investigating individual household members' time-space couplings when performing activities in projects. Various kinds of division of labor in households are identified from a large set of diaries written by spouses. Activity patterns are identified among the household members with various demands for energy. Two different ways to *share* activities in household organization projects are identified, and also three different ways to *divide* activities between the household members in joint projects. The organization of the daily activities and the timing of working hours strongly influences whether people share or divide activities in their projects. The research also indicates the consequences of policies regarding energy-saving advice.

In a similar vein, Palm and Ellegård (2011) discuss whom to target in policies on energy savings in homes. Social science research on energy use in the household sector rarely differentiates between the individual and the household. However, when formulating policy recommendations on how to reduce energy use in the home, it is vital to target the people who actually perform the activities concerned. Women perform most of the home-based household chores that demand electricity (for example, washing, dishwashing, vacuuming and cooking) at the same time as the electricity bill from the energy utility is most often sent to the man, who holds the household's contract. If, in such a case, information about how to save energy in the home is sent to the man, who does not perform these electricity-demanding chores, it is less likely that the information will affect the energy use. Instead, to achieve the desired effects of the information, it should be directed to, or at least made available to, the household member performing the most electricity-demanding household chores.

Aggregate analysis – recognizing the indivisible individual

The time-geographic approach requires of an aggregate description and analysis that it should be possible for the researcher to alternate between micro (individual) and macro (aggregates of various size, like the household or a sub-group of the whole sample of diarists) without losing important information when moving between the levels, as pointed out by Hägerstrand (1974).

Ellegård and Vrotsou (2006) developed visualization software (VISUAL-TimePAcTS)[14] for visualizing large numbers of time-geographic diaries on different levels of aggregation (individual, household, age group, gender, and so on, up to the whole population). The method, which takes inspiration from the thinking behind the principle presented in Figure 3.6, was further developed in Vrotsou (2010). The software is used in studies of the explorative kind, giving deeper insights into when, for how long and by whom daily activities are performed, at the chosen level of aggregation. It is possible to choose the individual household, group or population levels in the software.

Individuals' daily activity sequences at aggregate levels are visualized and demonstrate, for example, when in the course of the day activities of different kinds dominate among the individuals in the population. It is obvious that the individual diaries used to visualize the time use for activities in a population need to be ordered in a way that makes it possible to compare different visualizations. In VISUAL-TimePAcTS, the default order of the individual diaries is according to gender and age. The age and gender activity patterns appear at once in Figure 7.7, which shows two activities, cooking and eating, from diaries by a population of 463 individuals. Looking in more detail (see the lower part of Figure 7.7), it is obvious that women cook much more than men do. The upper part of Figure 7.7 shows that everybody eats. Lunch is eaten in a designated hour around midday, while dinners, or evening meals, are spread over a larger part of the late afternoon/evening, when people do not have so many other activities that they have to be involved in.

The aggregate activity patterns visualized by the program can be further analyzed with clustering and sequencing methods; such a methodology is presented by Hellgren (2015), who studied energy use as a consequence of activity sequences in his dissertation in technology and social change (see Chapter 8).

Vrotsou et al. (2017) present a further development based on the aggregate time-geographical visualization principle. The study is based on registers of individuals with severe mental illness in the Stockholm region, and consists of about 450 persons. The visualization is used to explore, in a 10-years perspective, what kinds of contacts (called *interventions* in the figure) these people diagnosed with severe mental illness have with the medical and social care sectors (institutions) – see Figure 7.8: the upper part shows men and the lower part women.

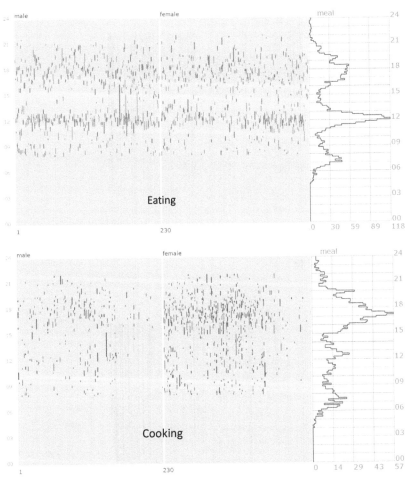

Figure 7.7 Cooking and eating activities in a population. In the figure the individuals (463 individuals aged 10 and up) are ordered according to gender (men to the left and women to the right) and age (the youngest to the right and the oldest to the left within each gender section). The lower part of the figure shows the spread of cooking activities and the upper part eating activities. The frequency graphs show how many people perform the same activity at each specific moment in the course of the day (in the frequency graph the indivisibility of the individual is collapsed).

The visualization software is interactive, and has three levels of detail in terms of interventions. It provides opportunities to explore various subgroups and also to focus on shorter time periods. This kind of visualization provides new knowledge about the lives of people with severe mental illness and interventions by the institutional care system they are subject to. Coupling and authority constraints

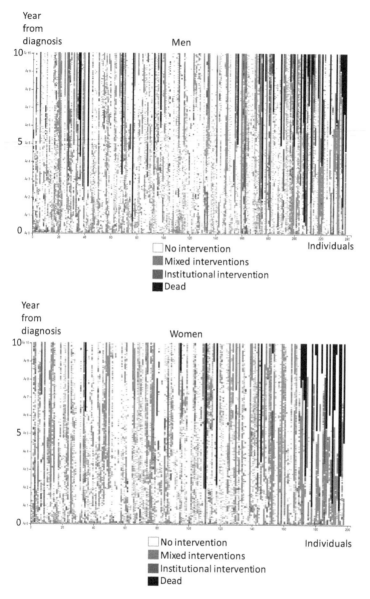

Figure 7.8 Interventions in the lives of people with severe mental illness in the first 10 years after diagnosis. Individual paths visualize each individual and the types of intervention that are directed to her. The upper part includes men and the lower part women. Within both gender categories the oldest people are located to the right and the youngest to the left. Black parts of the individual paths indicate that the person has died, white parts show no intervention and other nuances indicate various kinds of interventions. Year 0 in the visualization is when the specific person had the diagnosis, and what calendar year this was differs between the individuals. The figure gives an overview of the intensity of institutional interventions in their life; it also indicates when no interventions are done (white).

in their everyday life follow their illness and what kind of intervention they experienced. The visualization method is interactive, so researchers can look at the interventions at the very overarching level, as in Figure 7.8, but can also dig down into details of different groups of individuals and use different time scales. The categorization of the interventions is hierarchical, and so the many different kinds of interventions (ranging from prison to getting a phone call) can be identified at the individual level.

The applications of time-geographic thinking on everyday life issues often use time-geographic diaries and the constraint concepts. The process of writing the time-geographic diary is in itself helpful for the diarist, since they start thinking about what they do on a daily basis. This is a good point of departure for discussions with professional caregivers, since both parties in the discussion then start at a more informed level. Thereby, thoughts are nurtured about how activities might be changed; for example, by giving rewarding projects priority over projects that relate to bad feelings.

The results also might provide a groundwork for policy changes. Organization of care and services for people that are vulnerable should be arranged so that these people feel trust and security, and this poses demands on the service-providing organizations.

Hägerstrand worried about problems emerging from scientific specialization, even though he acknowledged the many positive effects of the scientific progress and technological development. He said: "The single person and the circumstances in her whole, indivisible life-world have disappeared between the masks in the web by which we have tried to capture the real world" (Hägerstrand 1977: 190, my translation).

The examples in this chapter show that it is possible to get an understanding of the whole, indivisible life-world from the use of the time-geographic approach.

Notes

1 Like bodily status and emotions.
2 See Chapter 6 regarding work on the assembly line.
3 Compare the now-zone in the subjective left part of Figure 3.7, which is different from the objectivistic now-line to the right in the same figure.
4 The project "Vardagslivets komposition" was funded by Riksbankens Jubileumsfond and lasted from 1989 to 1992.
5 The latter strive for average time use for average individuals and the indivisibility of the individual is not considered. See the discussion on added and average time use in Chapter 2.
6 For example, it became obvious that "leaving home" and "coming home" are important activities in people's lives since they mention them in the diaries and then, of course, they should be in the categorization scheme. Another important finding from the empirical bottom-up generation of activity categories was the importance of "doing nothing special". This takes time and so it should be regarded as an activity, even if it is ambiguous.
7 When all levels are included, there are about 600 activity categories in the time-geographical scheme.
8 There can also be other contexts for aspects of information that are included in the study.

9 The very first program was not accessible for the research community. In the 2000s free software was developed by Ellegård and Nordell, called Daily Life 2011 (*Vardagen* 2011).
10 It is presented in a later section of this chapter.
11 Several different apps have been developed to collect diary data. One of the most ambitious is found in the the research institute TOR (Tempus Omnia Relvelat) at Vrije Universiteit Brussel (www.vub.ac.be/TOR/). This is, however, not based on the time-geographic approach and applies other categorization schemes.
12 This is more of a traditional diary, wherein the women write about their thoughts, memories and hopes in a mix with experiences and valuations of events passing. The reflective diaries are exponents of the daily life perspective on time.
13 For example, besides the material layout of buildings and classrooms, sometimes technical aids were lacking and there were even examples of teachers who did not want to use a microphone, which meant that a hearing aid did not work for the student.
14 VISUAL-TimePAcTS is an abbreviation for Visualization (VISUAL), Time, Places (P), Activities (Ac), Technologies used (T) and Social companionship (S).

References

Andersson, G. 2009. Vardagsliv och boendestöd. En studie om människor med psykiska funktionshinder. Institutionen för Socialt arbete. Stockholms universitet. Diss.

Åström, E. 2009. Att lära, att gör, att klara. Förmedling av datortekniska hjälpmedel till barn med synnedsättning. Från förskrivning till vardaglig användning i skola och hem. Linköping Studies in Arts and Science, No. 487. Linköping University. Diss.

Bredland, E., Magnus, E., and Vik, K. 2015. Physical activity patterns in older men. *Physical & Occupational Therapy in Geriatrics*, Vol. 33, No. 1, pp. 87–102.

Buttimer, A. 1976. Grasping the dynamism of lifeworld. *Annals of the Association of American Geographers*, Vol. 66, No. 2, pp. 277–292.

Carlstein, T., Lenntorp, B., and Mårtensson, S. 1970. Individers dygnsbanor I några hushåll. *Urbaniseringsprocessen*, rapport 17. Institutionen för kulturgeografi och ekonomisk geografi, Lunds Universitet.

Carlstein, T. 1982. *Time, resources, society and ecology: on the capacities for human interaction in space and time in preindustrial societies*. Lund Studies in Geography, Series B Human Geography, No. 49. Lund, Sweden: C.W.K. Gleerup.

Cullen, I. 1972. Space, time and the disruption of behavior in cities. *Environment and Planning*, 4, pp. 459–470.

Dijst, M. 2014. Social connectedness: a growing challenge for sustainable cities. *Asian Geographer*, Vol. 31, No. 2, pp. 175–182.

Ellegård, K. 1993. Olikadant. Aspekter på tidsanvändningens mångfald. *Occasional Papers* 1993:4. Kulturgeografiska institutionen, Göteborgs universitet.

Ellegård, K. 1994. Att fånga det förgängliga. Utveckling av en metod för studier av vardagslivets skeenden. Vardagslivets komposition delrapport 2. *Occasional Papers* 1994:1. Kulturgeografiska institutionen, Göteborgs universitet.

Ellegård, K. 1999. A time-geographical approach to the study of everyday life of individuals – a challenge of complexity. *GeoJournal*, Vol. 48, No. 3, pp. 167–175.

Ellegård, K. 2006. The power of categorisation in the study of everyday life. *Journal of Occupational Science*, Vol. 13, No. 1, pp. 37–48.

Ellegård, K., and Friberg, T. 993. *Tiden bara rinner förbi*. Konsumentverket rapport 1992/93:27. Vällingby, Sweden: Konsumentverket.

Ellegård, K., and Nordell, K. 1997. *Att byta vanmakt mot egenmakt: Metodbok. Självreflektion och förändringsarbete i rehabiliteringsprocessen.* Stockholm, Sweden: Johansson & Skyttmo Förlag.

Ellegård, K., and Nordell, K. 2011. *Daily Life 2011.* Linköping, Sweden: Linköping University.

Ellegård, K., and Palm, J. 2015. Who is behaving? Consequences for energy policy of concept confusion. *Energies,* Vol. 8, No. 8, pp. 7618–7637.

Ellegård, K., and Vrotsou, K. 2006. Capturing patterns of everyday life – presentation of the visualization method VISUAL-TimePAcTS. Paper presented at the IATUR–XXVIII Annual Conference, Copenhagen, Denmark, 16–18 August 2006.

Erlandsson, L.-K. 2003. 101 women's patterns of daily occupations: characteristics and relationships to health and well-being. PhD diss., Lund University.

Eurostat. 2009. Harmonised European time use surveys 2008 guidelines. *Eurostat Methodologies.*

Hägerstrand, T. 1970. What about people in regional science? *Regional Science Association Papers,* Vol. XXIV, pp. 7–21.

Hägerstrand, T. 1974. Tidsgeografisk beskrivning. Syfte och postulat. *Svensk Geografisk Årsbok,* 1974, pp. 87–94

Hägerstrand, T. 1977. Att skapa sammanhang i människans värld – problemet. In *Om tidens vidd och tingens ordning. Texter av Torsten Hägerstrand.* G. Carlestam and B. Sollbe (eds). Byggforskningsrådet. Rapport T 21:1991.

Hellgren, M. 2015. Energy use as a consequence of everyday life. PhD diss. Linköping University.

Isaksson, C., and Ellegård, K. 2015. Dividing or sharing? A time-geographical examination of eating, labour and energy consumption in Sweden. *Energy Research & Social Science,* 10, pp. 180–191.

Journal of Occupational Science. 2006. Vol. 13, No. 1.

Kjellman, C. 2003. *Ta plats eller få plats? Studier av marginaliserade människors förändrade vardagsliv.* PhD diss., Lund University, 2003. English title: *To seize or to be given a place? Studies of marginalized people's changes in daily life.*

Kroksmark, U., and Nordell, K. 2001. Adolescence: the age of opportunities and obstacles for students with low vision in Sweden. *Journal of Visual Impairment & Blindness,* Vol. 95, No. 4, pp. 213–125.

Lenntorp, B. 1970. PESASP – en modell för beräkning av individbanor. *Urbaniserings processen,* rapport 38. Institutionen för kulturgeografi och ekonomisk geografi. Lunds Universitet.

Lenntorp, B. 1976. *Paths in space-time environments: a time-geographic study of movement possibilities of individuals.* Meddelanden från Lunds universitets Geografiska institutioner. Diss. LXXVII.

Magnus, E. 2009. *Student som alle andre: En studie av hverdagslivet til studenter med nedsatt funksjonsevne.* PhD diss. Norges Teknisk-Naturvitenskapelige Universitet.

Mårtensson, S. 1970. Tidsgeografisk beskrivning av stationsstruktur. *Urbaniseringprocessen,* rapport 39. Institutionen för kulturgeografi och ekonomisk geografi. Lunds Universitet.

Mårtensson, S. 1979. *On the formation of biographies in space-time environments.* Meddelanden från Lunds universitets Geografiska institution, Diss. LXXXIV, Lund.

Miller, R. 1982. Household activity patterns in nineteenth-century suburbs: a time-geographic exploration. *Annals of the Association of American Geographers,* Vol. 72, No. 3, pp. 355–371.

Nordell, K. 2002. *Kvinnors hälsa – En fråga om medvetenhet, möjligheter och makt. Att öka förståelsen för människors livssammanhang genom tidsgeografisk analys.* PhD diss. Gothenburg University.

Orban, K. 2013. *The process of change in patterns of daily occupations among parents of children with obesity – time use, family characteristics and factors related to change.* PhD diss. Lund University.

Orban, K., Edberg, A.-K., and Erlandsson, L.-K. 2012. Using a time-geographical diary method in order to facilitate reflections on changes in patterns of daily occupations. *Scandinavian Journal of Occupational Therapy*, 19, pp. 249–259.

Palm, R., and Pred, A. 1974. *A time-geographic perspective on problems of inequality for women.* Institute of Urban and Regional Development Working Paper 236. Berkeley, CA: Institute of Urban and Regional Development.

Palm, J., and Ellegård, K. 2011. Visualizing energy consumption activities as a tool for making everyday life more sustainable. *Applied Energy*, Vol. 88, No. 5, pp. 1920–1926.

Pred, A. 1981. Production, family and free-time projects: a time-geographic perspective on the individual and societal change in nineteenth-century U.S. Cities. *Journal of Historical Geography*, Vol. 7, No. 1, pp. 33–36.

Statistics Sweden (SCB). 2012. Nu för tiden. En undersökning om svenska folkets tidsanvändning år 2010/11. Levnadsförhållanden. *Rapport* 123.

Sunnqvist, C. 2009. Life events, stress and coping suicidal patients in a time-perspective. PhD diss. Lund University.

Sunnqvist, C., Persson, U., Westrin, Å., Träskman-Bendz, L., and Lenntorp, B. 2013. Grasping the dynamics of suicidal behaviour: combining time-geographic life charting and COPE ratings. *Journal of Psychiatric & Mental Health Nursing*, Vol. 20, No. 4, pp. 336–344.

Vrotsou, K. 2010. *Everyday mining: exploring sequences in event-based data.* PhD diss. Linköping University.

Vrotsou, K., Bergqvist, M., Cooper, M., and Ellegård, K. 2014. PODD: a portable diary data collection system. In *Proceedings of the 2014 International Working Conference on Advanced Visual Interfaces (AVI 2014), Como, Italy.* New York, NY: ACM, pp. 381–382.

Vrotsou, K., Andersson, G., Ellegård, K., Stefansson, C.-G., Topor, A., Denhov, A., and Bulow, P. 2017. A time-geographic approach for visualizing the paths of intervention for persons with severe mental illness. *Geografiska Annaler, Series B Human Geography*, Vol. 99, No. 4, pp. 331–359.

Westermark, Å. 2003. *Informal livelihoods: women's biographies and reflections about everyday life: a time-geographic analysis in urban Colombia.* PhD diss. University of Gothenburg.

8 Ecological sustainability – time-geographic studies on resource use

The time-geographic approach and environmental problems

Globally, ordinary people as well as politicians and decision makers in public and private organizations are concerned about the creeping of climate change and overuse of limited resources. Systems and constructions decided upon by industrialists and politicians long ago gave rise to the coal-driven infrastructures in society, subsequently complemented by the petroleum-dependent ones. Both facilitated more convenient heating and transportation systems and increasing productivity and efficiency in industrial production. Today, renewable energies stand on the threshold of a society that still hosts many old systems and infrastructures.

Decisions taken by those in power concerning what energy regime to lean on will induce myriad actions among people and organizations concerning, for example, what means of transportation to use, heating system to invest in, goods to purchase and projects to pursue. The mere volume of such actions affects both humans and the environment. In combination with people's increasing demand for goods and services, the growing infrastructural constructions and increased production of artifacts in the modern society result in a never-before-witnessed worldwide shuffling around of raw materials, products and people. In 1993, Torsten Hägerstrand said: "It is more and more clear that mankind has become a geological agent" (Hägerstrand 1993b: 161).[1] Hägerstrand was deeply concerned about the environmental problems caused by human action. For him, the development of the time-geographic approach was a way to pay attention to, describe, analyze and communicate the inevitable time-space couplings between humans and other phenomena in the environment. One concern of his was the increasing scientific specialization; he worried about it hiding mechanisms behind emerging environmental problems that are of an interdisciplinary character.

Layer upon layer of human action causes many environmental problems over time. What, then, is needed from humans in terms of changed resource use and reorganization of activities to put an end to, or at least mitigate, these destructive tendencies? There is a need for increased understanding of the effects of all the myriad human actions and the use of resources arising from them. Every industrial, commercial and private activity is performed somewhere by humans, as in a factory, shopping center or a home. Consequently, this is also where concrete actions to counteract the effects can be taken. However, there are structural restrictions, including authority constraints and physical barriers caused by earlier decisions and projects that bring a considerable inertia into the change process. Hägerstrand (1993a) labeled the landscape, which is both a host for the historic remains of dominant projects and a scene for current activities, the *processual* landscape.[2]

Towards an ecological approach

Hägerstrand's interest in ecological issues plays well with his critical view of scientific specialization. In the mid-20th century a split took place in the geography discipline, into physical geography and human geography. Researchers in the two parts of the discipline became less familiar with the works of each other. Hägerstrand wrote about the important task to "restore the links and reestablish a balance between the biophysical and the human branches of geography which are now mostly carrying on their business widely separated from each other" (Hägerstrand, 1976: 330). His time-geographic approach, with its cross-disciplinary orientation, laid the groundwork for investigations into how human projects affect both material surroundings and people's sustenance and wellbeing, and can help to bridge the cleavages within geography, between scientific disciplines and between research and sectors in society.

Policy makers and politicians make up plans to handle ecological problems, and if decided upon the plans are slowly implemented and the outcome in the physical world is usually limited. These problems inspired Hägerstrand to further the time-geographical approach and make efforts to introduce it not only into urban and regional planning (as demonstrated in Chapter 4), but also in the planning for a balanced use of natural resources (Hägerstrand 1988). During his long relationship with the Swedish national and regional authorities and policy makers, Hägerstrand identified a mismatch between what is planned to happen and the complex societal and material context in which the plans are to be implemented. He wrote that

> The material world within human reach is not altered by words but by the grips of the hand. The word-makers are in power, but for their decisions to turn into something more than vibrations in the air, one, some or all people must engage in the material things.
>
> (Hägerstrand 2009: 27, my translation)

Hägerstrand's message was that it is important to develop plans for creating structures that facilitate careful resource use. However, to affect the development, such plans to change structures must be put into practice. Once materially constructed and organized, they can serve as facilitators for people and organizations to act with care for the environment (Hägerstrand 1995). Even though the changes in resource use of each single person or organization in isolation do not make a big difference, taken together the amount of the many activities constitutes a force for change in itself (Hägerstrand 1989).

Decisions on how to design orders that enable and encourage people and their organizations to perform activities with less use of limited resources require that decision makers have concise knowledge about what governs individuals' everyday lives. People live in different contexts and thereby structural changes influence them differently. Everybody is constrained by their limited time and their inevitable spatial location, as pointed out by Hägerstrand (1970, 1976). Depending on what projects people want to, or must, pursue, and how they try to achieve the goals of these projects, people will meet authority, capability and coupling constraints, which will influence their opportunities to perform everyday activities without contributing to overexploitation of the resource base.

In a popular science article, Hägerstrand (1961) elaborates on historic property management and the results of the owner's decisions about changes in the structural conditions concerning utilization of the local resource base. The starting point is the long-term development of the management and activities of the old Svaneholm manor, located in Skåne in the southern part of Sweden. It was established in the 14th century, with large areas of land, including four villages. In the 18th century Rutger Maclean, a radical nobleman who modernized the agricultural and related activities, inherited it. He changed the authority constraints by initiating reform and reorganizing the farming system. In the new system the farmers were less dependent of the manor, and it resulted in increased productivity. The basic structure of this reformation system was later introduced in other places in Sweden to improve productivity in the agricultural sector. Hägerstrand reflects in general on work organization and utilization of local resources in farms of that time. He couples this development to those tracks that still are visible in the farming landscape. From the Svaneholm example, Hägerstrand shows that remains of historic human activities in the current landscape inform people about the organization of preindustrial farming activities.

Taking a 200-year perspective, Hägerstrand investigates what powers change the Swedish cultural landscape (Hägerstrand 1988). Based on his time-geographic worldview, he demonstrates the influence of laws decided upon centuries ago on the shape and content of today's landscape. The landscape is looked upon from different levels, from a satellite and a walking person to the biologist with his eye on the small animals on the ground. He shows that decisions from long ago lie hidden in what is visible now. For example, what kind of trees there are in a forest is decided upon by someone who has been advised (or even forced) by an authority on what to grow; the large fields of today are the results of decisions on land zoning, favorable economic conditions for investment in agricultural technologies and the dominance of large-scale farms with few people but many machines; and there are traces of the regulation of water (lake creation and drainage systems). The national decisions and the local farmers' decisions work together with nature's own forces in the creation of the landscape. From the long-term perspective, Hägerstrand makes it clear that the decisions taken by each forest owner will have repercussions for generations to come.

The Svaneholm example and the historical development of land use in Sweden as an outcome of resource regulation (authority constraints) give a picture of how "Every change in land use, constructions and buildings adds a new feature to the Earth's face" (Hägerstrand 1961, quoted in Carlestam and Sollbe 1991: 25, my translation).

Hägerstrand (1976) relates to the sustainability debate and calls for integration in geographical studies, and argues that human society is part of the pattern in the tapestry of nature woven by history. Taken together, this is a forerunner to what he later formulated as the *processual landscape* concept (Hägerstrand 1993a). The "processual landscape" is a conceptual tool for exploring human influences in a long-term perspective of the landscape. It concerns overarching societal policies, as well as individual-level activities. Here, Hägerstrand underlines the need for an ecological perspective, calling for policies and institutions to bridge the knowledge gap between sectors responsible for different aspects of the landscape.

In his final book, Hägerstrand (2009) did not actually use the concept of time-geography, even though he explains the basic concepts of time-geography and relates them to each other. He says that his approach is "all-ecological", based on the foundational idea about time-space couplings between individuals of different kinds in any geographically confined area.

Activities, land use and resources

In the modern, horizontally linked, long-distance society, the fundamental dependence on nature for humans' livelihoods is masked. For example, and as indicated in Chapter 1, all kinds of eatable items can be purchased and people do not need to know anything about the production chain of food. However, even if it is a progressive development that people no longer have to produce their own food from the grain and animals, it is problematic when the knowledge and insights about humans' dependence on nature, or rather about humans as part of nature, are limited or even disappear. Hägerstrand's time-geography is integrative in the respect that it underlines the couplings between (natural and human-made) resources on one hand and human life on the other. This was thoroughly researched by Tommy Carlstein, one of the members of Hägerstrand's research group.

Carlstein (1982) took on the challenge to investigate preindustrial societies and tried out the time-geographic concepts and the notation system in his investigations of ecological conditions circumscribing the life of people in these societies. Empirically, he used data from earlier published studies, mainly written by anthropologists. He searched for the logic behind the individual paths he identified from the time-space movements of the people described in the studies. This underlines the efforts people in the preindustrial societies made to handle the necessary couplings between household activities for sustenance and use of resources and land areas over time. Carlstein showed how coupling constraints impinge on people's activities, and emphasized the implications of the structure of technological and organizational systems on resource utilization. He also showed how such structures develop. Human organization of resource use, like that of water, cattle, crops and the various kinds of land use (e.g. grazing, growing crops, fallow), is closely related both to the season and the time of day. Sometimes the land areas in control of a group of nomads are located very far geographically from each other and people had to cross long distances to sustain themselves even in this kind of vertically linked society. Carlstein shows that the logic behind the movements for long-term sustenance relates to the utilization of the available resources, time and space. He identified two principles concerning how people use time and place, and used the concept *packing*. One principle concerns temporal-intense packing of land use and the other concerns spatial packing (density), and the two can be combined depending on, for example, climate, environment, land availability and people's time for cultivation. He discussed their different social organization and movements in the time-space (see Figure 5.2).

Less surprising, long-distance movements appear in the modern Swedish large-scale farming organizational projects, as shown by Solbär (2014). In order to run the large-scale farms, ruled by the industrialized society's principles of efficiency and productivity (which means fewer people and more machines), the farms' land use grows thanks to the farmers increasing the land area they control for farming. Then, of course, other farms close. However, from this enlargement process arises a problem of distance between the scattered fields of arable land. The farms close by may not be for sale so the farmer has to find land further away from the core farm's location. Consequently, the land in the farmer's control is not consolidated into one coherent piece, but fragmented into several fields, which are located some distance apart. Under these conditions, the result is that the farmers have to transport themselves (or their employees have to move), plus their machines, animals and products over long distances. In the day-long perspective they develop a nomadic travel pattern, which will affect their involvement in activities close to home.

Solbär (2017) had another take on the development of the modern agricultural landscape in her study on modern land reclamation. She points at a phenomenon in agriculturally dominated landscapes that might not be expected in a modern society, namely farmers' land reclamation projects. She regards the farm and its land as a pocket of local order, and shows how the farmer reshapes woodland into fields for growing crops or for grazing land, thereby creating a new order and layout of the pocket of local order. This reshaping is motivated by the need for the big farms to become what she labels more "time-space effective". The new order of the pocket of local order makes it possible for the farmer to use the time better and to intensify the utilization of the expensive machines. For example, the ploughing is more effective when the tractor does not have to go around a piece of woodland or move between two separate fields. The land reclamation results in larger fields, which decreases the need for spending time on driving the tractor between them. The bigger the farm, the more important is the organization of the couplings in the time-space between the people, the land and the machines. However, it also influences the space available for wildlife.

In a very different research field, but inspired by time-geography, Grow (2012) made an archaeologic comparative study of activities and resource use during the Late Neolithic Age and the Early Copper Age. The changing relationships between prehistoric human groups and the dynamic landscape were reanalyzed by utilizing the time-geographic concepts and notation system on the data. The path and prism concepts were useful for reconstructing activities and shifts in the archaeological context.

Social organization, resource use and human livelihoods in a Laotian village in transition towards modernization were investigated by the geographers Nishimura et al. (2010). Data on household members' daily activities were collected by GPS technology and interviews. The transition process from the vertically linked society, with total dependence on natural resources in sustenance projects, to entering a horizontally linked society by becoming partly dependent on wage earnings from employed work in an urban setting, is illustrated by the fundamentally different activities performed by different household members. The transition creates new conditions for the division of labor among the household members, which is exemplified by their daily movements in the time-space. The area within which the household member working in the nearby town makes her daily moves increases and because of the work schedule she is away from the home most of the time she is awake. Consequently, she cannot participate in other activities in the household's organizational projects for sustenance, like fishing, collecting insects or selling at the market.

The Japanese geographer Kushiya (1985) made one of the first empirically grounded applications of time-geography to human behavior in relation to natural resources. In this study, the couplings intertwining human activity with natural resources are put to the fore. He studied how fishermen used the waters of Tokyo Bay as a resource and adjusted their activities according to their knowledge about exactly when and where the catch is located under different seasonal and weather conditions. Hence, the daily activities of fishermen are heavily influenced by vocational knowledge about the time-space location of the fish. Even though only the article's abstract is in English, its time-geographic visualizations give a good foundation to interpret the activities performed. They reveal the potential both of the notation system as a nonverbal language and of time-geography as an integrative tool when studying human use of resources, both aimed at by Hägerstrand.

Anderberg (1996) did a system-policy-level study from a time-geographical perspective wherein he critically evaluated traditional flow analyses, using empirical data of material flows in the Rhine Basin. The societal and natural perspectives are linked to each other and Hägerstrand's concept of processual landscape is applied in the critical examination of conventional flow analyses.

In a study of the development of the dairy industry in Sweden, Ellegård (1977) shows that when the authorities in the early 20th century prohibited dairies from running co-located pig farms to get rid of surplus products from the production process (skimmed milk and whey), the dairies instead let those by-products out with the waste water. The result was eutrophication of rivers and lakes. This shows how an authority constraint set by national decision makers with the intention of improving hygiene in dairy production had severe repercussions for the landscape. It was overcome after some years by the invention of technology to dry the by-products and sell the dried products on the market as food for calves and pigs.

Sanglert (2013) used time-geography to investigate how historical heritages are cared for and maintained in the modern Swedish landscape. He uses Hägerstrand's processual landscape concept to analyze what happens at a specific place over a long time span. A heritage place is somewhere designed long ago for a specific purpose, as a grave, arable land or a building, but now the remains mainly have a value as a place of cultural heritage, for materializing history to modern people and for tourism. Sanglert relates to what Hägerstrand acknowledged as a problem of colliding interest. Different administrative bodies have different ideas about how the land on which the heritage is located should be utilized, and their conflicting interests influence the future of the object. Sanglert suggests extended coordination of the diverse administrative organizations that share the responsibility for maintaining the heritage places in the landscape.

Resource use in buildings and homes

Citizens in modern societies live in buildings that, depending on the outdoor temperature, are heated or cooled, since people want a convenient indoor climate. This is one component of the energy use in the household sector; another one concerns the use of electricity for manifold household appliances. In several countries in Europe the energy use for the household sector is about 40% of the total national energy use, and since decreasing energy use is one important issue in the efforts to mitigate climate change it is vital to know more about how households may contribute (Swedish Energy Agency 2015).

Lenntorp (1993) investigated the supply of material for constructing homes in the Nordic countries. Utilizing the time-geographical tools and concepts he identified the complex chains of production regarding defining components of houses (like windows, foundations, the roof and the chimney). He also mapped the materials used for the preproduction of such components, which took place in factories located elsewhere, and the necessary transportation of components to the building site. The strong economic incitements for centralized production in the construction industry encourage centralized production of materials and components in large-scale factories far away from the building site. The outcome is long-distance transportation and corresponding high use of fossil fuel and increasing demand for transportation in the horizontally linked society. It also calls for expansion of the road infrastructure. This will facilitate more transportation.

Glad (2006) investigated the spread and implementation of the concept of passive houses in the Swedish construction industry from the late 1990s. Passive houses should not use any purchased energy for heating. Instead, heat is generated from the activities of the human bodies staying in the house, the sun and people's use of appliances. The building shell, of course, must be very tight. Glad shows how the innovative concept was transferred from one building site to another, initially mainly promoted by one architect. Together with European colleagues, he acted as a missionary in the construction industry and among policy makers. Construction companies increasingly implement the passive house concept, both for single-family houses and apartment buildings. An increasing awareness among policy makers at the European level about the importance of reducing energy use in the housing sector to mitigate climate change took the concept of the passive house (the terms *low-energy houses* and *zero-energy houses* were used later) into energy policies and regulations for the building sector.

There are interdisciplinary research groups in Sweden performing studies on household members' energy use in homes, generating time-geographically-grounded knowledge about energy use as derived from individuals' activities in their daily lives. The results of such research are suitable for engineering research to develop better simulation models regarding energy use in buildings (Widén et al. 2009, 2012).

In a study of households living in apartments in multi-family passive houses, Karresand (2013) dives deep into single households' daily use of electrical appliances, and uses the time-geographic approach to understand their associated use of electricity. She develops the *energy order* concept to capture people's activities related to household projects. She found that people's ways of ordering their activities for doing laundry, cooking and entertainment vary, but there are some similarities in their approaches depending on their couplings to other household members, and scheduling of other activities they are involved in. Electricity is embedded in daily life activities, and time-geographic concepts and tools help explain why it is hard to see where energy savings can be made, and also how to find out and pay attention to such savings in the activities when pursuing projects. Karresand's detailed study of the household members' use of appliances for their projects shows that even with identical appliance settings, activities are performed differently, which results in different levels of electricity use. For example, drying clothes in the laundry project can be done very differently and require more or less electricity. One energy order relates to drying clothes by hanging them to dry in the air. This demands little electricity, but it claims relatively more time and space, both for the hanging of each item and for the drying process. In contrast, there is another energy order for drying clothes, where an electric dryer is the technological base. It demands much more electricity, but the person doing the related activities uses little time. Karresand shows that these household projects are influenced by coupling, capacity and authority constraints, which influence the household members' opportunities to prioritize energy savings in creating their energy orders.

Hellgren (2015) is inspired by the time-geographical VISUAL-TimePAcTS visualization tool and combines it with cluster analysis to show the total energy use among people in Sweden at the aggregate level. In his study, the definition of energy use for the household sector deviates from that used in the national statistics (Swedish Energy Agency 2015). Hellgren includes both households' energy use for transportation and domestic electricity use in the total usage, which adds a considerable amount to the energy used by the households. Based on data from a national time-use survey performed by Statistics Sweden 2010/2011, he takes the individuals' daily activity sequences as a point of departure to calculate the energy use, both in the home (from cooking, laundry and cleaning activities) and for transportation (going by car, public transit, air and boat). Hellgren clusters the individuals according to their energy use as it is generated from the activity sequences, and identifies five clusters. The electricity use varies depending on the time of the day and in all clusters except one the energy use for transport is higher than that of domestic electricity. Consequently, the widened definition of household-sector energy use gives important background knowledge for inspiring households to make changes. He concludes that energy-saving information should be differently designed and formulated depending on what cluster of activity pattern the individuals targeted belong to.

The environmental concerns are a dominating strand in the thinking and writings of Hägerstrand. The long-term outcome of processes at work in the landscape, be they results of human activities like new technologies or nature's own adjustments to new conditions, is that

> most Swedes now live in a "room of technology" (the city plus the traffic system). The biologically dominated landscape is no longer something that the majority of the population put their hands into or even have as a workplace. This landscape has merely become a backdrop, which is rapidly passed by the traveler, or literally jumped over by air travelers.
>
> (Hägerstrand 1988: 52, my translation)

What, then, will happen with the development of knowledge about the foundation from which humans live? Hägerstrand's contribution is to show the immense source of knowledge that lies in what happened in the past. His investigation of the development of population distribution, forestry and land use in the light of policy and administrative decisions and regulations (Hägerstrand 1988) gives inspiration to new takes on understanding what people's activities mean in the processual landscape.

Notes

1 The final result is labeled *Anthropocene* (Castree 2015), an era where human activities generate effects that change fundamental conditions of the Earth system.
2 Hägerstrand was not satisfied with the opportunities to express what he meant when he constructed the concept *förloppslandskap* in Swedish. He said that *förlopp* is richer, content-wise. He regarded the German concept *Verlaufslandschaft* closer to what he wanted to say than the English *process*.

References

Anderberg, S. 1996. Flödesanalys i den hållbara utvecklingens tjänst – Reflektioner kring en 'metabolism'-studie av Rhenområdets utveckling. PhD diss. Lund University.

Carlestam, G., and Sollbe, B. 1991. *Om tidens vidd och tingens ordning*. Texter av Torsten Hägerstrand. Byggforskningsrådet. Rapport T21:1991.

Carlstein, T. 1982. *Time resources, society and ecology. On the capacity for human interaction in space and time in preindustrial societies*. The Royal University of Lund, Department of Geography. Lund Studies in Geography, Series B Human Geography, No. 49. Lund, Sweden: C.W.K. Gleerup.

Castree, N. 2015. The Anthropocene: a primer for geographers. *Geography*, 100, Part 2, p. 66.

Ellegård, K. 1977. Utveckling av transportmönster vid förändrad teknik – en tidsgeografisk studie. Mimeo. Forskargruppen i kulturgeografisk process- och systemanalys. Lunds universitets kulturgeografiska institution.

Glad, W. 2006. *Aktiviteter för passivhus. En innovations omformning i byggprocesser för energisnåla bostadshus*. Linköping Studies in Arts and Science, 367. Linköping University. Diss.

Grow, K. 2012. Time geography: a reanalysis of spatial shift in the Great Hungarian Plain Institute for European and Mediterranean Archaeology. *Chronika*, 2, pp. 66–74.

Hägerstrand, T. 1961. Utsikt från Svaneholm. *Svenska Turistföreningens årsskrift*, pp. 33–64.

Hägerstrand, T. 1970. What about people in regional science? *Regional Science Association Papers*, Vol XXIV, pp. 7–21.

Hägerstrand, T. 1976. Geography and the study of interaction between nature and society. *Geoforum*, 7, pp. 329–344.

Hägerstrand, T. 1988. Krafter som format det svenska kulturlandskapet. *Mark och vatten år 2010*. Stockholm, Sweden: Bostadsdepartementet, pp. 16–55.

Hägerstrand, T. 1989. Globalt och lokalt. *Svensk geografisk Årsbok*, 65, pp. 9–19.

Hägerstrand, T. 1993a. Samhälle och natur. *NordREFO*, 1, pp. 14–59.

Hägerstrand, T. 1993b. What about nature in regional science? In *Visions and strategies in European integration: a North European perspective*. L. Lundqvist and L.O. Persson (eds). Berlin, Germany: Springer-Verlag, pp. 155–161.

Hägerstrand, T. 1995. A look at the political geography of environmental management. *LLASS* Working Paper No. 17.

Hägerstrand, T. 2009. *Tillvaroväven*. K. Ellegård and U. Svedin (eds). Stockholm, Sweden: Formas.

Hellgren, M. 2015. *Energy use as a consequence of everyday life*. PhD diss. Linköping University.

Karresand, H. 2013. Creating new energy orders: restrictions and opportunities for energy efficient behaviour. In *Conference Proceedings 2013 ECEEE Summer Study – Rethink, Renew, Restart*, 3–8 June 2013, Belambra Les Criques, Presqu'île de Giens, Toulon/Hyères, France, pp. 2147–2158.

Kushiya, K. 1985. Time-geographic interpretation of fishermen's daily activities on Tokyo Bay, Japan. *Geographical Review of Japan*, Series A, Chirigaku Hyoron 58, pp. 645–662.

Lenntorp, B. 1993. De fyra nordiska husen. *NordREFO* 1993:1.

Nishimura, Y., Okamoto, K., and Boulibam, S. 2010. Time-geographic analysis on natural resource use in a village of the Vientianne Plain. *Southeast Asian Studies*, 47, pp. 426–450.

Sanglert, C.-J. 2013. Att skapa plats och göra rum - Landskapsperspektiv på det historiska värdets betydelse och funktion i svensk planering och miljövård. Institutionen för kulturgeografi och ekonomisk geografi, Lunds universitet. Diss.

Solbär, L. 2014. *Anthropogenic open land in boreal landscapes. Investigations into the creation and maintenance of arable fields on Swedish farms.* Faculty of Social Sciences and Department of Human Geography. Lund University. Diss.

Solbär, L. 2017. Gårdsdriften och arronderingen – den nutida nyodlingens utgångspunkter. *Bebyggelsehistorisk Tidskrift*, 2017, Issue 73, pp. 47–63.

Swedish Energy Agency. 2015. *Energiläget.*

Widén, J., Lundh, M., Vassileva, I., Dahlquist, E., Ellegård, K., and Wäckelgård, E. 2009. Constructing load profiles for household electricity and hot water from time-use data-modelling approach and validation. *Energy and Buildings*, Vol. 41, No. 7, pp. 753–768.

Widén, J., Molin, A., and Ellegård, K. (2012). Models of domestic occupancy, activities and energy use based on time-use data: deterministic and stochastic approaches with application to various building-related simulations. *Journal of Building Performance Simulation*, Vol. 5, No. 1, pp. 27–44.

Part III
Spread, criticism and future

9 International spread and criticism

Paving the way for time-geography

As shown in previous chapters, the thinking behind what was later presented as the time-geographic approach gradually emerged during a long research process starting before the mid-20th century, when Torsten Hägerstrand did empirical work in Asby, first on migration (Hägerstrand 1950), and then for his PhD thesis on diffusion of innovations (Hägerstrand 1953/1967). The general objective of his thesis was *a problem*, the process of innovation diffusion, rather than a geographical region.[1] (Hägerstrand 1953/1967). However, even though this problem is general, empirical investigations were necessary for evidential reasons. Hägerstrand followed the geographical spread of innovations over time, based on the idea that human individuals are carriers of innovations and as they use the novel phenomena and communicate with their neighbors, there is an opportunity for the innovation to spread.[2] However, all people who were exposed to the innovation did not accept the novelty and Hägerstrand used the Monte Carlo method to simulate which individuals took it up. The general approach, then, concerned innovation diffusion as it is carried by people in a continuous social process of change in the time-space. This was something fundamentally different from looking at the mere geographic distribution of the innovation over the population in a region at different points in time (see Figure 2.3). With its quantitative approach, the simulation model in itself was innovative in human geography and yielded considerable interest in the international geographic society.[3] It was a radical break from the regional geography orientation, which by then was the dominating tradition in human geography[4] (Hägerstrand 1983). The influence of his thesis is indicated by the fact that even though it was written in Swedish, English-speaking geographers with knowledge of Swedish recognized its novelty. It was reviewed in an international journal in 1954 (Lenntorp 2008).

The combination of Hägerstrand's earlier work on migration chains (Hägerstrand 1957; see Chapter 2) and his initial thoughts on the innovation diffusion model (Hägerstrand 1953/1967) laid the intellectual groundwork for the time-geographic approach, which was consequently developed generically out of the experiences and results of the earlier research efforts and problems. There is a seed for this development in the introduction of his PhD thesis, where Hägerstrand writes:

Existence in a society implies that people are constantly in motion. Almost every individual possesses his unique movement figurations, with his residence at the center, and with places of work, shops, places of recreation, residences of intimate friends, and other similar locales as nodal points.

(Hägerstrand 1967: 8;[5] in Swedish 1953: 13–14)

This later became a core point in the time-geographic approach: the indivisible individual in the time-space.

Time-geography as modernity

However, it was not until the late 1960s that Hägerstrand found his thoughts ordered coherently enough to be presented as the time-geographic approach with its concepts and notation system. At a seminar in the geography department at Lund University, he gave a presentation of his ideas, and the interest was great among his colleagues in Lund and elsewhere in Sweden. Hägerstrand gave the opening speech at the 9th Conference of the Regional Science Association in Copenhagen, Denmark, where geographers, economists and sociologists interested in regional development met. His speech was published in 1970 (Hägerstrand 1970), marking the international breakthrough of time-geography. Soon the new way of thinking, its concepts and the originality of the visualizations based on the notation system gave rise to invitations asking Hägerstrand to give presentations at scientific meetings, conferences and seminars, not only in Sweden but also in other countries; for example, the USA, UK, Canada, England and Holland.[6]

Time-geography became influential, and many geographers were proud of it being an original geographical contribution (Thrift 1977, 2005; Pred 1978, 1981, 1983, 1984). It eventually had some influence in other disciplines; for example, the sociologist Giddens recognized its innovative way of handling time and space in his structuration theory (Giddens 1984).

Because of Hägerstrand's inspirational research in many research fields, the geography department in Lund became a hot spot in human geography and researchers from all over the world went there. The international geographic society talked about "the Lund School" and "the Hägerstrand School", and because of the width of Hägerstrand's research engagements it is hard to say whether this label concerned the migration, the diffusion of innovation or the time-geography orientation. Hence, in the 1970s visiting researchers had different scientific motives to go to Lund for discussions with Hägerstrand. For example, the Japanese professor Teruo Ishimizu, who spent some months in Lund, was mainly motivated by Hägerstrand's use of quantitative methods. However, when Ishimizu went back to Japan he introduced the time-geographic individual path and prism concepts in a geography reader in Japanese, which was one of the first sources of inspiration for young Japanese geography students (Okamoto and Arai 2019).

Some guests in the 1970s were interested in getting deeper insights in what the new time-geographic approach was about and wanted to present it internationally, like the human geographers Allan Pred from the USA and Nigel Thrift from England, and the Canadian sociologist William Michelson. Most visitors, including the geographers Peter Hagget, Bill Mead and Anne Buttimer, and the economist Andrew Harvey, were intrigued by the intellectual environment developing at the geography department and wanted to be involved in discussions with Hägerstrand and his colleagues.

Allan Pred stayed in Lund during the 1960s as a guest researcher and played an important role for the international spread of Hägerstrand's innovation diffusion research with its novel use of quantitative methods.[7] This was a vital contribution to the human geographic discipline, which in the mid-20th century was searching for legitimacy in science. In addition, Pred was guest editor for a special issue of *Economic Geography*[8] in 1977, where some original articles in time-geography were published. The theme of the issue was planning-related Swedish geographic research, and it served as an introduction to the time-geographic approach for many geographers worldwide.

Inspired by the articles and by Professor Ishimizu's visit to Lund, a research group on time-geography was initiated at Tokyo University by Professor Arai. The Japanese time-geography research group translated some original time-geography articles into Japanese (Arai et al. 1989). In its own research this group focused on the household division of labor, timing of work and opportunities to travel – all constituents of importance for living a good everyday life (Okamoto et al. 2019). One young PhD student from China, Yanwei Chai, got inspiration from this group and upon returning to China he introduced time-geography by the turn of the century. Chai is now Professor in Geography at Peking University (Zhang et al. 2019).

The international interest in time-geography was boosted by the publication of three edited volumes on the theme of timing space and spacing time by Carlstein[9] et al. (1978), which contributed to the discussion about time and space in social theory. Volume 2 was dedicated to time-geography and comprised several articles written by members of Hägerstrand's research group in Lund.

Time-geography in citations

Citations of publications in a scientific field is one way to get an overall idea about the spread of ideas in the international literature. Citations of Hägerstrand's works may serve as a brief indicator of his impact in the international scientific society over time. Persson and Ellegård found three distinct clusters of citations of Hägerstrand, each one thematically centered around one of his core publications in the research orientations discussed above (Persson and Ellegård 2012).[10] The first cluster appearing in the citation analysis concerns citations of Hägerstrand (1957), which is his first publication on migration in English. The second cluster to appear concerns research on the geographic diffusion of innovations and refers to the English translation of his PhD thesis (Hägerstrand 1953/1967). The third and final cluster concerns time-geography and its use in analyzing activity, travel and space. These citations are centered on the first publication in English where the time-geographic approach was presented: Hägerstrand (1970).

The publication of the speech given by Hägerstrand at the meeting of the Regional Science Association (Hägerstrand 1970) yielded huge international interest and is still the most cited publication by Hägerstrand.[11] There was a peak in citations of Hägerstrand's works in the mid-1980s, after which the freshness and modernity of his works faded. Hägerstrand had retired and his influence in the geographical sphere decreased, even though he went on publishing a lot, now on time-geography with an ecological orientation. By the end of the 1990s, and partly as a result of research by a new generation of geographers in the USA, interest in time-geography took off again, and citations of Hägerstrand's works peaked once again. The young researchers H.J. Miller and M.-P. Kwan made substantial contributions to the development of time-geography and utilized the improved computer power and new software. Kwan is now a professor at the University of Illinois and Miller is a professor at the Ohio State University.

After the death of Hägerstrand in 2004, a never-before-seen number of citations of his works appeared.

Time-geography criticized

Critical discussion and reflection over new approaches are important drivers of scientific development and it is a sign of health in the academic debate when eager acceptance of new approaches meets opposition. When presented around 1970, the time-geographic approach was not a fully-fledged theory; it was an outline of the approach, which since then has been developed and employed in many contexts. The previous chapters have indicated the range of themes where time-geography has served, beyond the initial examples.

In the 1970s, there was a wave of intense criticism towards the novel time-geographic approach. The critics claimed that time-geography was too materialistic and did not take human feelings and experiences into consideration (Buttimer 1976; Baker 1979). Such criticism has occasionally appeared over the years since then (Dijst 2019; Rose 1993). There were also critical voices involving concerns over the way time and power is handled in time-geography (Rose 1977; Baker 1979).

The time-geographic notation system with its individual path was used in many early publications to visualize the problems studied (Hägerstrand 1970; Lenntorp 1976; Carlstein 1982; Miller 1982; Thrift 1977; Palm and Pred 1974). The central idea of the time-geographical notation system is to visualize the movements of individuals in the time-space, and the main tool for doing so is the individual path (see chapters 2 and 3). The individual path shows when an individual has been located where in the time-space in the past. For the future location of the individual, the prism was developed. The prism shows the opportunity space in which an individual can choose to pave her way in her efforts to achieve the goals of the projects she is involved in, given the constraints in her local context. Researchers in social science criticized time-geography for focusing the material, Outer, world (the time-space movements of the individual) shown by the individual path and for not acknowledging people's experiences and feelings (Thrift and Pred 1981; Buttimer 1976).

This critique is based on the assumption that the individual concept equals a human being, which is not always the case in time-geography (see Chapter 2).

Hägerstrand used the individual concept in a more general sense (see Chapter 3). In one article, Hägerstrand presented the idea of landscape evolution as a system in which each individual performs a kind of ballet in the time-space (Hägerstrand 1976).[12] Here, he uses the wide time-geographic definition of the individual concept, which includes humans as well as man-made and natural things (see Chapter 2). In such a ballet, individual paths could, in principle, be used for illustrating the movements of each of the many different kinds of individuals "on stage" in the landscape, consequently not only humans. Anne Buttimer, a geographer of ideas and a phenomenologist, was both intrigued and challenged by the time-geographic approach. She was especially critical of the individual path and the physicality in its performance. In an article from the mid-1970s she said that the movements of the individual paths in the time-space, or the "ballet" in the landscape, were nothing but a "dance macabre", a skeleton of the underlying intentions and experiences of those involved (Buttimer 1976). However, despite her criticism, she paid extended visits to the geography department in Lund for discussions with Hägerstrand. Buttimer also started a joint research project with him, the Dialogue Project, wherein they videotaped interviews with established university professors about their biographies, including both their professional and private life. Thus, it was possible to identify a link between the individual path of each professor and his (there were very few *hers*) development of scientific thoughts, feelings, experiences and private life (Dialogue Project 1977–1985).[13]

In an article from 1977, Courtice Rose criticized the way time-geography handles the time dimension; he found the existential aspects of time in human life to be overlooked. Baker (1979) delivered another critique in the same vein. He said that time-geography fails to consider human agency, and argued that time-geography neglected the role of power and ideology for making better geography and argued that time-geography did not consider the ideological dimension.

Members of Hägerstrand's research group took up the criticism and both Lenntorp (1976) and Mårtensson (1979) discuss the tension between the Inner (subjective) and the Outer (physical material) worlds of the individual in their PhD theses. They each present a figure (Lenntorp 1976: 14; Mårtensson 1979: 151; see also Hägerstrand 2009: 229) where this is put to the fore.[14] These figures resemble each other, and indicate how to deal with human individuals' feelings, experiences and will in combination with the physical appearance of the individual's movements as illustrated by the individual path and the constraints met in the daily activity context (compare Figure 3.7). However, their dealing with this criticism did not play any significant role in the international debate about time-geography as a materialistic approach.

As mentioned in Chapter 4, Hägerstrand had a dual role, both as a researcher and as an expert in governmental and public investigations. Pred (1977) argued in one article on time-geography that, given Hägerstrand's close relationship with the reformation of the welfare state and municipalities in Sweden, there is a danger that time-geography might be "mistakenly construed as nothing more than a planning tool" (Pred 1977: 213). Pred's fear did not gain ground and Hägerstrand himself was critical of the authorities that did not utilize his suggestions (Öberg 2005).

In the early 1990s, time-geography met criticism concerning gender issues. Gillian Rose (1993) said that time-geographic research missed the specific feminine experiences, and characterized time-geography as "social science masculinity" (Rose 1993: 40). This interesting statement reveals an assumption that time-geography as a whole equals the works of one man, Hägerstrand. However, the mere composition of Hägerstrand's research group says something else. In the mid-1960s, when Hägerstrand had funding and could recruit research assistants, he employed one woman and two men. By that time it was unusual for women to be engaged in research projects in Sweden. In the 1970s more women got involved in the group. Hägerstrand reflected upon the fact that so many women were attracted by the time-geographic approach. He thought that it might be because it highlights everyday, mundane business that women perform on a daily basis. However, Rose's criticism was more nuanced than the single citation above indicates, and she also wrote that "time geography shares the feminist interest in the quotidian paths traced by people, and again like feminism, links such paths, by thinking about constraints, to the larger structures of society" (Rose 1993: 18).

Criticizing time-geography for having a masculine bias is somewhat problematic since much of the very early time-geographic research, actually concerned the situation of women.[15] Even if inequality between men and women was not the main focus, several time-geographic works at least opened a way of analyzing the constraints that influence men and women differently in the same society and time-geography was used to bring the household division of labor into focus (Miller 1982; Palm and Pred 1974; Hägerstrand and Lenntorp 1974; Mårtensson 1979). Somewhat later, Kwan (1999) dealt with gender inequality in daily travel related to errands and activities in the home, and Friberg (1990) showed gender differences in daily life from a female perspective.

The interest in time-geography grew during the 1990s, a development related to the opportunities for researchers to utilize the increased capacity of computers. The new technical tools opened up the potential to simulate and calculate, for example, the prism (opportunity space with its corresponding potential path area) of individuals (Miller 1991) and to create visualizations with a huge number of individual paths from large sets of travel diaries (Kwan 1999). Later on, Kwan used multimethod approaches to investigate minority women's everyday life conditions in the USA (Kwan 2008).

In addition, the computer-aided time-geographic diary method was developed, and it eventually gained interest both in human geography and in occupational therapy (see Chapter 7). Based on the indivisible individual, such methods aimed to go deeper into analyses of the constitution of daily life. The diary method takes experiences and feelings into the time-geographical approach (for example, Ellegård 1999; Nordell 2002; Westermark 2003; Kroksmark et al. 2001) and most of the studies performed concern the everyday lives of women. Several occupational therapists noted that the idea of the indivisible individual as presented in the time-geographic approach fits well with their work with the rehabilitation processes of individuals (Magnus 2009; Kroksmark et al. 2006; Orban 2013; Bendixen and Ellegård 2014).

At the end of his final book, Hägerstrand underlines the doubleness of human existence. There is the physical existence of the material body in the Outer world (Popper's World 1) and there is the Inner world (Popper's World 2), and in the latter there is what living individuals

> carry with them inside, their models of their worlds, behavioral programs and human intentions and projects. These are always present in the fabric of existence and provide a context to every now by a cross-section from the invisible past and future. Everybody is in contact with her private world, but only experience, gained knowledge and the content of current communication may give rise to ideas about what might happen – what the implications are – for actors in the neighborhood and at longer distance.
>
> (Hägerstrand 2009: 271, my translation)

Notes

1 In the 1940s and 1950s geography was dominated by regional geography, describing regions from dominant properties.
2 In Hägerstrand (1986) he illustrated this and showed the difference in societies with more or less speedy means of transportation.
3 A long debate was going on among geographers regarding the usefulness of quantitative methods and regional descriptions respectively. See, for example, Schaefer (1953) and Hartshorne (1958).
4 Lenntorp (2008) puts Hägerstrand's PhD thesis into its scientific context and discusses the consequences of it being written in Swedish and not translated until 14 years after its defense.
5 In Swedish: "Samhällslivet förutsätter, att människorna är i ständig rörelse. Nästan varje individ torde äga sin speciella rörelsefigur med bostaden i centrum och med arbetsplatser, affärer, fritidslokaler, umgängesvännernas bostäder och dylika ställen som nodpunkter" (Hägerstrand 1953, pp. 13–14).
6 This might be one reason for the scattered publication pattern of Hägerstrand (see Chapter 4). There was not much time to write longer texts and organizations engaging Hägerstrand were eager to publish what he had said at seminars and meetings.
7 He translated Hägerstrand's thesis into English 14 years after its defense (Hägerstrand 1967).
8 *Economic Geography*, 1977, Vol. 53, No. 2.
9 The editor, Carlstein, was one of the members of Hägerstrand's research group.
10 This article puts the spread of citations of Hägerstrand's works in the global time-space context.
11 It is still the most-cited article written by Hägerstrand with 3,808 citations reported on www.google.se/search?q=what+about+people+in+regionals+cience+citations&ie=utf-8&oe=utf-8&client=firefox-b&gfe_rd=cr&dcr=0&ei=TUHXWcXwJM6AtAHgv4Jo (searched 6 October 2017).
12 Compare also the citation in the beginning of this chapter, from the thesis (Hägerstrand 1953/1967, p. 8).
13 The Nobel Museum in Stockholm produced a "time-space aquarium" showing the movements in time-space of the Nobel laureates in economics. This representation

revealed the importance of a location (Chicago) to which researchers had moved from various remote places in the world before getting the award.
14 Compare figures 3.8 and 3.9, which are inspired by these two figures and by a similar one in Hägerstrand (2009).
15 Also, since the early 1990s "time-geography days" have been arranged in Sweden on a yearly basis. Interestingly, the majority of both the organizers and participants are female researchers. During the last five to 10 years, more men have joined in the arrangements.

References

Arai, Y., Kawaguchi, T., Okamoto, K., and Kamiya, H. (eds). 1989. *Seikatsu no Kuukan, Toshi no Jikan* (*Anthology of Time-geography*). Tokyo: Kokon-Shoin.

Baker, A. 1979. Historical geography – a new beginning? *Progress in Human Geography*, 3, pp. 560–570.

Bendixen, H.-J., and Ellegård, K. 2014. Occupational therapists' job satisfaction in a changing hospital organisation: a time-geography-based study. *Work*, Vol. 47, No. 2, pp. 159–171. doi: 10.3233/WOR-121572.

Buttimer, A. 1976. Grasping the dynamism of the lifeworld. *Annals of the American Association of Geographers*, Vol. 66, pp. 277–292.

Carlstein, T., Thrift, N., and Parkes, D. 1978. *Timing space and spacing time*. Vol. 2, *Human activity and time geography*. London: Edward Arnold.

Carlstein, T. 1982. Time resources, society and ecology. On the capacity for human inter-action in space and time in preindustrial societies. The Royal University of Lund, Department of Geography. *Lund Studies in Geography, Series B Human Geography*, No. 49. C.W.K. Gleerup.

Dialogue Project. 1977–1985. *Dialogue project*. Lund, Sweden: University Library Archive, Lund University.

Dijst, M. 2019. A relational interpretation of time-geography. In *Time-Geography in the Global Context*. K. Ellegård (ed.). Abingdon and New York, NY: Routledge.

Economic Geography. 1977. Planning-related Swedish geographic research. *Economic Geography*, Vol. 53, No. 2.

Ellegård, K. 1999. A time-geographical approach to the study of everyday life of individuals – a challenge of complexity. *GeoJournal*, Vol. 48, No. 3.

Friberg, T. 1990. *Kvinnors vardag. Om kvinnors arbete och liv. Anpassningsstrategier i tid och rum*. Meddelanden från Lunds Universitets geografiska institutioner, avhandlingar No 109.

Giddens, A. 1984. *The constitution of society. Outline of the Theory of Structuration*. Cambridge: Polity Press.

Hägerstrand, T. 1950. Torp och backstugor i 1800-talets Asby (1950). In *Från Sommabygd till Vätterstrand*. E. Hedkvist et al. (eds). Linköping, Sweden: Tranås hembygdsgille, pp. 30–38.

Hägerstrand, T. 1953/1967. *Innovationsförloppet ur korologisk synpunkt*. Lund, Sweden: Gleerupska Universitets-bokhandeln. Translated into English by Allan Pred as *Innovation Diffusion as a Spatial Process*. Lund, Sweden: C.W.K. Gleerup.

Hägerstrand, T. 1957. Migration and area. Survey of a sample of Swedish migration fields and hypothetical considerations on their genesis. In *Migration in Sweden: A Symposium*. D. Hannerberg, T. Hägerstrand and B. Odeving (eds). Lund Studies in Geography, Series B Human Geography, No. 13. Lund, Sweden: C.W.K. Gleerup, pp. 27–158.

Hägerstrand, T. 1970. What about people in regional science? *Regional Science Association Papers*, Vol XXIV, pp. 7–21.

Hägerstrand, T. 1976. Geography and the study of interaction between nature and society. *Geoforum*, 7, pp. 329–344.

Hägerstrand, T. 1983. In search for the sources of concepts. In *The Practice of Geography*. A. Buttimer (ed.). Harlow: Longman Higher Education, pp. 238–256.

Hägerstrand, T. 1986. Tiden och tidsgeografin. Tidens vidd och tingens ordning. Några synpunkter på innovationsförkopplets historiska geografi. In *Om tidens vidd och tingens ordning. Texter av Torsten Hägerstrand. Byggforskningsrådet*. G. Carlestam and B. Sollbe (eds). Rapport T21:1991.

Hägerstrand, T. 2009. *Tillvaroväven*. K. Ellegård and U. Svedin (eds). Stockholm, Sweden: Formas.

Hägerstrand, T., and Lenntorp, B. 1974. Samhällsorganisation i tidsgeografiskt perspektiv. In *Bilagedel 1 till Orter i regional samverkan*. Statens Offentliga Utredningar (SOU) 1974: 2. Stockholm: Arbetsmarknadsdepartementet, pp. 221–232.

Hartshorne, R. (1958) The concept of geography as a science of space, from Kant and Humboldt to Hettner. *Annals of the Association of American Geographers*, Vol. 48, No. 2, pp. 97–108.

Kroksmark, U., and Nordell, K. 2001. Adolescence: the age of opportunities and obstacles for students with low vision in Sweden. *Journal of Visual Impairment & Blindness*, Vol. 95, No. 4, pp. 213–125.

Kroksmark, U., Nordell, K., Bendixen, H.J., Magnus, E., Jakobsen, K., and Alsaker, S. 2006. Time geographic method: application to studying patterns of occupation in different contexts. *Journal of Occupational Science*, Vol. 13, No. 1, pp. 11–16.

Kwan, M.-P. 1999. Gender, the home–work link, and space-time patterns of nonemployment activities. *Economic Geography*, Vol. 75, No. 4, pp. 370–394.

Kwan, Mei-Po. 2008. From oral histories to visual narratives: re-presenting the post-September 11 experiences of Muslim women in the USA. *Social & Cultural Geography*, Vol. 9, No. 6, pp. 653–669.

Lenntorp, B. 1976. *Paths in space-time environments. A time-geographic study of movement possibilities of individuals*. Meddelanden från Lunds universitets Geografiska institutioner. Diss. LXXVII.

Lenntorp, B. 2008. Innovation diffusion as spatial process (1953): Torsten Hägerstrand. In *Key Texts in Human Geography*. P. Hubbard, R. Kitchin and G. Valentine (eds). Thousand Oaks, CA: Sage Publications, pp. 1–8.

Magnus, Eva. 2009. *Student som alle andre: En studie av hverdagslivet til studenter med nedsatt funksjonsevne*. PhD diss., Norges Teknisk-Naturvitenskapelige Universitet.

Miller, Harvey J. 1991. Modelling accessibility using space-time prism concepts within geographical information systems. *International Journal of Geographical Information Systems*, Vol. 5, No. 3, pp. 287–301.

Miller, R. 1982. Household activity patterns in nineteenth-century suburbs: a time-geographic exploration. *Annals of the Association of American Geographers*, Vol. 72, No. 3, pp. 355–371.

Mårtensson, S. 1979. *On the formation of biographies in space-time environments*. Meddelanden från Lunds universitets Geografiska institution, Diss. LXXXIV, Lund.

Nordell, K. 2002. *Kvinnors hälsa – En fråga om medvetenhet, möjligheter och makt. Att öka förståelsen för människors livssammanhang genom tidsgeografisk analys*. PhD diss., Gothenburg University.

Öberg, S. (2005). Hägerstrand and the remaking of Sweden. *Progress in Human Geography*, Vol. 29, No. 3, pp. 340–349.

Okamoto, K., and Arai, Y. 2019. Time-geography in Japan – its application to urban life. In *Time-Geography in the Global Context*. K. Ellegård (ed). Abingdon and New York, NY: Routledge.

Orban, K. 2013. *The process of change in patterns of daily occupations among parents of children with obesity – time use, family characteristics and factors related to change*. PhD diss., Lund University.

Palm, R., and Pred, A. 1974. *A time-geographic perspective on problems of inequality for women*. Institute of Urban and Regional Development Working Paper 236. Berkeley, CA: Institute of Urban and Regional Development.

Persson, O., and Ellegård, K. 2012. Torsten Hägerstrand in the citation time-web. *Professional Geographer*, Vol. 64, No. 2, pp. 250–261.

Pred, A. 1977. The choreography of existence: comments on Hägerstrand's time-geography and its usefulness. *Economic Geography*, Vol. 53, No. 2, pp. 207–221.

Pred, A. 1978. The impact of technological and institutional innovations on life content: some time-geographic observations. *Geographical Analysis*, Vol. 10, No. 4, pp. 345–372.

Pred, A. 1981. Production, family and free-time projects: a time-geographic perspective on the individual and societal change in nineteenth-century U.S. cities. *Journal of Historical Geography*, Vol. 7, No. 1, pp. 33–36.

Pred, A. 1983. Structuration and place: on the becoming of sense of place and structure of feel. *Journal of the Theory of Social Behavior*, Vol. 13, No. 1, pp. 45–68.

Pred, A. 1984. Place as historically contingent process: structuration and the time-geography of becoming places. *Annals of the American Geographers*, Vol. 74, No. 2, pp. 279–297.

Rose, C. 1977. Reflections on the notion of time incorporated in Hägerstrand's time-geographic model of society. *Tijdschrift voor Econ. en Soc. Geografie*, Vol. 68, No. 1.

Rose, G. 1993. *Feminism and geography: the limits of geographical knowledge*. Cambridge: Polity Press.

Schaefer, F. 1953. Exceptionalism in geography: a methodological examination. *Annals of the Association of Geographers*, Vol. 43, No. 3.

Thrift, N. 1977. *An introduction to time-geography*. Concepts and Techniques in Modern Geography 13. Norwich: Geo Abstracts.

Thrift, N. 2005. Torsten Hägerstrand and social theory. *Progress in Human Geography*, Vol. 29, No. 3, pp. 337–340.

Thrift, N., and Pred, A. 1981. Time-geography: a new beginning. *Progress in Human Geography*, 5, pp. 277–286.

Westermark, Å. 2003. *Informal livelihoods: women's biographies and reflections about everyday life: a time-geographic analysis in urban Colombia*. PhD diss., University of Gothenburg.

Zhang, Y., Chai, Y., and Tan, Y. 2019. The time-geographic approach in research on urban China's transition. In *Time-Geography in the Global Context*. K. Ellegård (ed.). Abingdon and New York, NY: Routledge.

10 Time-geography – from the past into the future

Time-geography as a mode of thinking

A Swedish colleague of Hägerstrand, Håkan Törnebohm, a professor in the theory of science, made a distinction between compositional and contextual syntheses (Törnebohm 1972). In a compositional synthesis the parts of a whole are hierarchically mapped and organized, while the contextual synthesis regards how the properties of an object relate to the various contexts in which it appears. Hägerstrand (1974) argued that the regional geography tradition can be regarded as a compositional synthesis, while his time-geographic approach contributes to the development of a contextual synthesis in human geography.

The relationships between objects' properties and the contexts wherein these objects appear can be described as couplings. Time-geography underlines the importance of couplings in the time-space for individuals' successful pursuit of their projects. Hägerstrand did not find any adequate vocabulary for the various forms of time-space couplings he identified. This motivated his development of the time-geographic concepts and notation system in which the time-space couplings (and de-couplings) can be clearly identified. Without such a notation it is difficult to describe and analyze the many aspects of the couplings between individuals in the time-space, not to mention the difficulties of communicating in a precise way about them.

The time-geographic notation system, then, may serve as a basis for contextual synthesis identifying the complex weaving of time-space movements and couplings between individuals caused by their intentions and efforts to be located in the same place as other important individuals (which includes resources) in time for getting in touch to achieve goals. Hägerstrand (1974) said that when using the notation system "It shall be easy to understand the phenomenon in the real world that is illustrated by the visualization" (Hägerstrand 1974: 88, my translation) This relates to the objectivistic side of time-geography with its anchoring in the material world, and implies that the representation should be transparent and invite deeper reflection and discussion about the phenomena visualized. Not only physical phenomena per se can be visualized, but also ideas and cultures as they appear from their material outcomes. However, for an analysis of ideas and culture, the notation system serves as a clarifying basic point and should be used in combination with theories suitable for the problem.

Hägerstrand shows how the interrelated time-geographical concepts (individual path, prism, population and project) can be utilized to interpret what eventually appear in specific time-space locations, pockets of local order (Hägerstrand 1985). Such a pocket is defined as a delimited part of the time-space, which for that time period and at that place is dominated by an order (furnishing, layout and things, and social agreements) created from human intentions, ideas and negotiations. This acknowledges the subjective world's influence on what is realized in the material, Outer world. Once ideas are materialized and the rules and agreements are temporarily stabilized by the individuals' performing activities in the pocket, the order is accepted and it will imbue most further actions. The pocket of local order concept also brings analysis of power to the fore and the concept bears a potential for investigations of arrangements of many kinds, from the micro to the macro scale (Lenntorp 1998). A room, a home, a building, a block, a city district and so on may serve as the geographical bases for studies of people's daily lives and their use of resources in different kinds of societies.

As is obvious from this book, visualizations play an important role in the time-geographic literature. The notation system is very different from other kinds of representations, not least depending on the strongly held assumption about the indivisible individual and the strict time-space dimensions along which the individual path is used to represent individuals' moves and other activities. Sometimes, however, the notation system is mistaken for the time-geographic approach as a whole, and time-geography is seen as solely materialistic. Hopefully, this book has provided arguments and evidence for the proposition that time-geography is much more than its notation system. At the same time, the notation system is a fundamental tool for getting deeper and more general insights about complex situations, where the individual paths illustrate couplings between individuals of different kinds, in the past as well as now. It shows the limits of their opportunities to couple in the future (the prism). The individual path concept also helps reveal what consequences arise for other individuals from one individual's decisions to act in a certain way. Hägerstrand said: "The visualization should generate questions which would not be posed without it" (1974: 88, my translation).

In search for the well-known but not recognized

Time-geography should serve as a means to reveal general principles, which are hidden in the manifold small couplings in the time-space (not least by their self-evident character). Human individuals have to rationalize their activities in their everyday existence in order not to drown in a never-ending flow of thoughts about what to choose from the existing alternatives. By experiencing repeated daily activity situations in childhood and adolescence, people learn how to master such common situations (the daily routines like morning toilet and dressing, the meal structure, and so on). These become internalized and taken for granted, and people stop reflecting over doing them; they just *do*. When such activities are internalized they become self-evident and only when put into question or performed in a strange way do they receive attention. The opportunity to pay attention to things that are self-evident is one important point of departure for time-geographic analysis of everyday life, especially when the aim is to change an activity performance that has become habitual. For example, a person with unhealthy eating habits may discover the non-conscious eating pattern from a time-geographic visualization, where all moments of eating in the course of the day are visualized in the context of the individual's full activity sequence. Not until they are paid attention to can such habits can be altered.

Time-geography is a way to illuminate neglected and non-conscious aspects of existence. Sometimes, however, a reaction to time-geographic presentations about people's everyday lives is that this is self-evident and, consequently, it is not important to research. This is the core point – the self-evident is simple and easy to refer to once it is made conscious and paid attention to, but it is not thought of and acted on in the moment when it happens in the situation when the person has to choose. What is experienced as self-evident when paid attention to must also be transformed into altered activities; otherwise no change will follow.

Time-geography and new technologies

The developments of information and communication technologies (ICT), with GPS cell phones, tablets and social media, and so on, have brought new and never-before-witnessed opportunities to collect huge amounts of data about movements in time and space. The ICT also influence daily activity sequences. What do they mean to time-geography?

Couclelis (2009) argued that because activities are increasingly performed digitally, researchers have hit the limits of traditional time-geography. She suggested a re-examination of the prism in light of the fact that ICT loosen the traditional links between activity, place and time, and presented a conceptual model of people's activities in a multidimensional space, with the power of traditional time-geography in mind. Also, Shaw and Yu (2009) argued for new concepts in time-geography due to the development of ICT. ICT-related activities are even said to collapse both time and space (Castells 2009; Cairncross 2001). This might be so from the perspective of ideas (compare Popper's World 2, the world of ideas; see Chapter 2), but still the human individual and her appliances are located somewhere in space and exist over time, and the movement possibilities of the individual with her body (in Popper's World 1, the physical world) are limited by the range of the prism. This is true even if the individual can communicate his ideas (in Popper's World 2) with people on the other side of the globe in (approximately) real time. The daily activity sequences are affected by increasing ICT use (Thulin et al. 2019) and new kinds of projects and activities appear. A person might, for example, shop over the internet, but the material objects purchased must still be transported from a production site to the home of the customer. The material existence of the body and things is inevitable. This example shows that the fundamental relevance of the time-geographic approach with its ontology, concepts and notation system will not be displaced.

When it comes to the huge amount of data produced by ICT, big data, another kind of problem, arises. Even if it is possible to get huge amounts of data from people's use of cell phones, the data quality has to be discussed. There are some methodological problems as well; for instance, the location of the phone is not by definition the same as the location of the person. Even more important are the ethical issues from researchers' use of big data collected by companies for other purposes. Schwanen (2016) raises questions of this type which should be recognized.

Time-geography as a tool for communication between disciplines

As a communicative tool, time-geography might increasingly serve as a bridge between disciplines. As mentioned in Chapter 9, a stream of criticisms towards time-geography concerns the subjective dimension. A number of the studies used in this book demonstrate how time-geography is combined with theories from disciplines where the subjective dimension is crucial; for example, studies in social work, psychiatry and occupational theory. This shows that time-geography is useful in research where subjective dimensions are vital. In addition, it shows that time-geography is a useful means to open up partitions between disciplines and a basis for communication about research problems.

The time-geographic approach, then, serves as a common basis of communication, supports communication between researchers in different theoretical traditions and facilitates mutually reflected understanding of research problems. In combination with theories from different scientific fields, time-geography helps in addressing the big challenges in society. Its tools and concepts might also assist in forming policies and regulations for the creation of structures which facilitate sustainable production and use of resources. Many societal changes are outcomes of the myriad human activities performed one by one by many individuals for a long time. This variety is a result of what it is possible for people to do within the dominating structures and orders of society. Since many unfavorable outcomes regarding, for instance, social and environmental problems are generated by many people doing activities in a way that is problematic at a societal and global level, it should be possible to generate favorable outcomes by creating orders and organizing material structures that guide people to act in a sustainable way.[1] In this endeavor, the time-geographic understanding of everyday life, where individuals are indivisible and subject not only to authority and capability constraints but also coupling constraints, will provide useful insights that otherwise would not arise.

In his final book, *The Fabric of Existence*, Hägerstrand wrote:

> There are two prerequisites for designing forms for human activity that will keep the global change within the frames of sustainable development. One concerns knowledge about what can be done and what should be avoided. The other concerns diffusion of the understanding of how to use this knowledge, and guarantees for these insights to be accepted in practical use.
>
> The most difficult thing is probably to encourage practical use of the knowledge. In that respect there are manifold diverse actors who must be mobilized in a coordinated way. Among them, there are surely many who sit on the fence.
>
> (Hägerstrand 2009: 23, my translation)

Time-geography will continuously be utilized by researchers in a variety of disciplines. But what turns might the development of the approach take? At least two possible future directions can be foreseen: one towards specializations and one towards contextualization of time-geographic research. The specialization direction could lead to time-geographical concepts and methods being used and further developed *within* various disciplines and research fields. This means that various kinds of time-geography might develop, but the interdisciplinary fertilization from the approach will diminish. The contextualization direction implies that the development of time-geography takes place within disciplines, but also that the researchers from different disciplines fertilize each other's work in open-minded discussions at conferences and in networks where time-geography as a whole is targeted.

Which direction the future developments of time-geography will take lies in the minds, hands and activities of researchers who use and creatively develop the approach, its methods and concepts. The proof of the usefulness of time-geography will be found in the utilization of its methods and concepts in studies of social and ecological problems in society. Of vital importance is approaching problems which are produced by the myriad activities performed when now constantly transforms future into past.

Note

1 A related example is the opportunities arising from the new ICT structure to created crowd-funding campaigns.

References

Cairncross, F. 2001. *The death of distance: how the communications revolution is changing our lives*. Brighton, MA: Harvard Business School Press.

Castells, M. 2009. *The rise of the network society – the information age, economy, society and culture, Vol. I*. 2nd ed. Malden, MA, and Oxford: Blackwell.

Couclelis, H. 2009. Rethinking time geography in the information age. *Environment and Planning A*, Vol. 41, No. 7, pp. 1556–1575.

Hägerstrand, T. 1974. Tidsgeografisk beskrivning: Syfte och postulat. *Svensk Geografisk Årsbok*, 50, pp. 86–94.

Hägerstrand, T. 1976. Geography and the study of interaction between nature and society. *Geoforum*, 7, pp. 329–344.

Hägerstrand, T. 1985. Time-geography: focus on the corporeality of man, society, and environment. In *The Science and Praxis of Complexity*. Tokyo, Japan: United Nations University, pp. 193–216.

Hägerstrand, T. 2009. *Tillvaroväven*. K. Ellegård and U. Svedin (eds). Stockholm, Sweden: Formas.

Lenntorp, B. 1998. Orientering I ett forskningslandskap. In *Svensk kulturgeografi. En exkursion inför 2000-talet*. M. Gren and P.-O. Hallin (eds). Lund, Sweden: Studentlitteratur, pp. 67–85.

Schwanen, T. 2016. Geographies of transport II: reconciling the general and the particular. *Progress in Human Geography*, Vol. 40, No. 1, pp. 126–137.

Shaw, S-L., and Yu, H. 2009. A GIS-based time-geographic approach of studying individual activities and interactions in a hybrid physical-virtual space. *Journal of Transport Geography*, Vol. 17, No. 2, pp. 141–149.

Thulin, E., and Vilhelmson, B. 2005. ICT-based activities among young people – user strategies in time and space. *Tijdschrift voor Economische en Sociale Geographie*, Vol. 96, No. 5, pp. 477–87.

Thulin, E., and Vilhelmson, B. 2019. Bringing the background to the fore: time-geography and the study of mobile ICTs in everyday life. In *Time-Geography in the Global Context*. K. Ellegård (ed.). Abingdon and New York, NY: Routledge.

Törnebohm, H. 1972. *Perspektiv på studier*. Avdelningen för Vetenskapsteori, Göteborgs Universitet, Rapport Nr. 51.

Index

Page numbers in *italics* refer to figures. Page numbers in **bold** refer to tables.